ACHIEVING EXCELLENCE IN BUSINESS

QUALITY AND RELIABILITY

A Series Edited by

EDWARD G. SCHILLING

Coordinating Editor

Center for Quality and Applied Statistics
Rochester Institute of Technology
Rochester, New York

W. GROVER BARNARD

*Associate Editor for
Human Factors*

Vita Mix Corporation
Cleveland, Ohio

RICHARD S. BINGHAM, JR.

*Associate Editor for
Quality Management*

Consultant
Brooksville, Florida

LARRY RABINOWITZ

*Associate Editor for
Statistical Methods*

College of William and Mary
Williamsburg, Virginia

THOMAS WITT

*Associate Editor for
Statistical Quality Control*

Rochester Institute of Technology
Rochester, New York

Additional volumes in preparation

Managing for World-Class Quality: A Primer for Executives and Managers, *Edwin Shecter*

Quality Engineering Handbook, *edited by Thomas Pyzdek and Roger W. Berger*

A Leader's Journey to Quality, *Dana M. Cound*

ACHIEVING EXCELLENCE IN BUSINESS

A Practical Guide to the Total Quality Transformation Process

Kenneth E. Ebel
Consultant
Chattanooga, Tennessee

ASQC Quality Press
Marcel Dekker, Inc.

Milwaukee

New York • *Basel* • *Hong Kong*

Library of Congress Cataloging-in-Publication Data

Ebel, Kenneth E.
   Achieving excellence in business : a practical guide to the total
quality transformation process / Kenneth E. Ebel.
      p.   cm. – (Quality and reliability ; 27)
   Includes bibliographical references and index.
   ISBN 0-8247-8522-3
   1. Total quality management.   2. Success in business.   I. Title.
II. Series.
HD62.15.E25   1991
658.5'62–dc20   91-20737
                                                        CIP

This book is printed on acid-free paper.

American Society for Quality Control
310 West Wisconsin Avenue
Milwaukee, Wisconsin 53203

Marcel Dekker, Inc.
270 Madison Avenue, New York, New York 10016

Current printing (last digit):
10 9 8 7 6 5 4 3 2 1

PRINTED IN THE UNITED STATES OF AMERICA

About the Series

The genesis of modern methods of quality and reliability will be found in a simple memo dated May 16, 1924, in which Walter A. Shewhart proposed the control chart for the analysis of inspection data. This led to a broadening of the concept of inspection from emphasis on detection and correction of defective material to control of quality through analysis and prevention of quality problems. Subsequent concern for product performance in the hands of the user stimulated development of the systems and techniques of reliability. Emphasis on the consumer as the ultimate judge of quality serves as the catalyst to bring about the integration of the methodology of quality with that of reliability. Thus, the innovations that came out of the control chart spawned a philosophy of control of quality and reliability that has come to include not only the methodology of the statistical sciences and engineering, but also the use of appropriate management methods together with various motivational procedures in a concerted effort dedicated to quality improvement.

This series is intended to provide a vehicle to foster interaction of the elements of the modern approach to quality, including statistical applications, quality and reliability engineering, management, and motivational aspects. It is a forum in which the subject matter of these various areas can be brought together to allow for effective integration of appropriate techniques. This will promote the true benefit of each, which can be achieved only through their interaction. In this sense, the whole of quality and reliability is greater than the sum of its parts, as each element augments the others.

The contributors to this series have been encouraged to discuss fundamental concepts as well as methodology, technology, and procedures at the leading edge of the discipline. Thus, new concepts are placed in proper perspective in these evolving disciplines. The series is intended for those in manufacturing, engineering, and marketing and management, as well as the consuming public, all of whom have an interest and stake in the improvement and maintenance of quality and reliability in the products and services that are the lifeblood of the economic system.

The modern approach to quality and reliability concerns excellence: excellence when the product is designed, excellence when the product is made, excellence as the product is used, and excellence throughout its lifetime. But excellence does not result without effort, and products and services of superior quality and reliability require an appropriate combination of statistical, engineering, management, and motivational effort. This effort can be directed for maximum benefit only in light of timely knowledge of approaches and methods that have been developed and are available in these areas of expertise. Within the volumes of this series, the reader will find the means to create, control, correct, and improve quality and reliability in ways that are cost effective, that enhance productivity, and that create a motivational atmosphere that is harmonious and constructive. It is dedicated to that end and to the readers whose study of quality and reliability will lead to greater understanding of their products, their processes, their workplaces, and themselves.

Edward G. Schilling

Preface

In order to be successful, a business must consistently offer products or services that

- Fulfill an actual need
- Satisfy customer's expectations
- Comply with applicable standards and specifications
- Comply with statutory and other requirements of society
- Are competitively priced
- Are provided at a cost which promotes long-term viability of the organization

What should we do to accomplish this? Who is responsible and who should be involved? How do we ensure consistency and assure that everyone is working together?

In the context of a specific enterprise, these questions can only be answered by those knowledgeable of the business, its personnel and customers, applicable laws and regulations, and quality precepts, systems, and techniques. Of these areas, quality precepts, systems, and techniques are not customarily within the realm of experience of most business managers.

This book was developed to assist managers in finding answers to these questions and in ensuring the long-term success of their business. You might say it provides a guide to learning and achieving true understanding of quality management. (As one of the most experienced contributors to this book noted, "You can't learn the quality process through shortcuts—you learn it as you learn a difficult sport: by experience, practice, failure and new tries!")

As such, this book identifies basic quality precepts and provides succinct guidance on the process of developing effective quality management and a total quality culture. It is based on the belief that the most effective course on quality management is self-guided, as it allows management to tailor learning to the organization's unique needs and integrate receipt of knowledge with experience.

The information in this book is presented so as to be applicable to all enterprises, including sole proprietorships, corporations, not-for-profit foundations, and government organizations. In each of these categories, the overall process for transforming the organization's culture and management systems is the same. The differences lie in the decisions made during the transformation, decisions which are based upon each organization's unique needs, culture, systems, and processes.

This book is a living guide to achieving excellence. It reflects the experiences and expertise of business and quality management professionals from a wide variety of industries (major contributors are recognized in the Acknowledgments). Ultimately, however, this goal can only be achieved with your help.

Please let me know of your tribulations and trials in implementing this book, and your ideas for improving this book. (Improvement is the keyword and ideas are its substance.) Send your comments and suggestions to Ken Ebel, 3101 North Causeway Blvd., Suite D, Metairie, LA 70002.

I hope that this book will lead you to the same successes as the organizations which are setting the quality pace. And in turn, you will help others pursue excellence through personal involvement and leadership, and by contributing to future revisions of this book.

Kenneth E. Ebel

Acknowledgments

The following people and organizations contributed their time and ideas to the creation of this book.

American Society for Quality Control (ASQC), Quality Press, Milwaukee, WI

ASQC Writing Group 28, "Service Quality Standard"

Bill Bornhoeft, CQE, CQA (an interested party in quality evolution), Chattanooga, TN

Clyde Brewer; Brewer & Associates, Inc., Grand Prairie, TX

J. Don Brock; President, ASTEC Industries, Inc., Chattanooga, TN

Ernie Coppola; Quality and Safety Manager, Commercial Nuclear Fuel Plant, Babcock & Wilcox Fuel Company, Lynchburg, VA

Joseph F. Decosimo, CPA; Senior Partner, Joseph Decosimo & Company, C.A.A., Chattanooga, TN

W. Leonard Fant; President, Hutcheson Medical Center, Fort Oglethorp, GA

David Rodney Fraley; Artist, Miamisburg, OH

Dr. James Grogan, Vice President, Medical Affairs, and Wanda Pool, Accreditation/Quality Management Coordinator; Erlanger Medical Center, Chattanooga, TN

Beverly Inman-Ebel, CCC-SLP; Director, Communication Clinic, Chattanooga, TN

Summerfield K. Johnston, Jr.; CEO, Johnston Coca-Cola Bottling Group, Inc., Chattanooga, TN

E. Paul Loch; Quality Practitioner, Westinghouse Electric Corporation, Orlando, FL

Olan Mills II, CEO, and Terry Blunt, Director of Human Resources; Olan Mills, Inc., Chattanooga, TN

P. Robert Philp; President, First Tennessee Bank, N.A., Chattanooga, TN

Douglas R. Rosensteel; Statistical Management Consultant, Saltsburg, PA

Robert C. Steffy, Jr.; Executive Vice President, Power, Tennessee Valley Authority, Chattanooga, TN

Joel D. Susman, CPA; Joel Susman and Associates; Chattanooga, TN

Russ Utlak; Total Quality, Commercial Nuclear Fuel Department, Westinghouse Electric Company, Monroville, PA

Dave West; Principal Senior Consultant, Quality & Productivity Associates, Chattanooga, TN

Clyde Brewer and Beverly Inman-Ebel are afforded special appreciation for the extensive personal and professional support which they provided throughout the creation of this book and for the detailed reviews which they performed. This book might never have been published if not for their efforts.

David Rodney Fraley is also afforded special appreciation for his artistic expressions which are used throughout this book. (As you will see, his expression of excellence is a healthy, vibrant tree.)

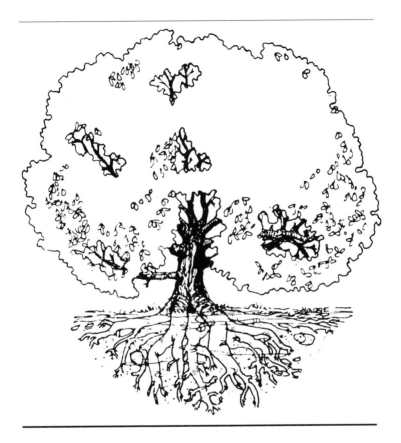

Contents

_____ CHAPTER 1 _____

Introduction

Quality is a subject which has been receiving increasing attention. This heightened awareness has produced a myriad of publications on the importance of quality and various aspects of quality systems and tools. Unfortunately, these publications tend to fall into two basic categories: promotion of awareness and detailed guidance. The first stimulates action based on general concern but without a full understanding of the tools, or leads the reader to believe that the answer lies in only one tool (this path is often wrought with failure). The second is generally written for implementation by very large organizations and tends to require interpretation and integration by quality specialists (which few small organizations know they can afford).

The combination of the heightened awareness and lack of concise guidance has lead to unrealistic expectations, lack of understanding regarding implementation, frustration and resignation by many managers, employees, and many members of the general public.

Many organizations are currently using some of the tools addressed in this book. However, excellence is not effectively pursued through the use of "a few tools." Excellence requires a systematic approach to developing a Total Quality culture and building Total Quality into the fabric of the organization, including:

- Transforming the attitudes of the people in the organization,
- Developing an integrated set of improvement mechanisms which provide for a never-ending organization-wide effort,
- Controlling the quality of products, services, and activities,
- Gearing up for strategic quality management.

This book seeks to encourage organizations of all shapes and sizes to manage the quality of their activities in a more effective manner and to succinctly provide a roadmap to the Total Quality transformation process. In essence it is a self-guided as-you-go course on quality management and a desk reference for managers who are leading or contributing to a transformation in their organization's culture, systems, and processes. It recognizes and is based upon the following:

- Consistently meeting and exceeding the customer's quality expectations is the precursor to consistent growth, while failure to meet the customer's quality expectations can have unforeseen consequences for both the customer and the organization,
- It costs substantially less to keep current customers than it does to replace them. (The vast majority of customers will go elsewhere without explanation or notice when their expectations are not met, and many will go elsewhere when their expectations are not exceeded.)
- Management has the responsibility to ensure customer satisfaction, which also means that management is responsible for preventing problems,
- People in a business form the principal resource of the business,
- Quality transformation is like a snowball rolling downhill (optimum effect occurs when it starts at the top),
- Effective management systems must be structured to provide for business growth and responsiveness to changing situations,
- Excellence is pursued only through continuous quality improvement,
- Improvement is dependent on the rate at which new knowledge is obtained and used,

- Understanding is achieved through the application of new knowledge, success, failure, and new tries.

This book is divided into four parts, each building upon the understanding gained through its predecessor. However, before delving into this book, please become familiar with the terminology presented in "Special Word Usages" at the end of this chapter. "Special Word Usages" is provided as an aid in interpreting the information in this book and minimizing confusion that arises from the multiplicity of definitions contained in English language dictionaries.

Part 1, "Quality Precepts and Quality System Principles," provides a general understanding of quality precepts, principles of quality management, and the quality system (Chapters 2 and 3).

Part 2, "Pathway to Excellence (Quality Culture and System Development Process)," provides a roadmap for nurturing a Total Quality culture, and developing and implementing a quality management system which

- Utilizes the talents and knowledge of all members of the organization, and
- Ensures that the efficiency and effectiveness of all aspects of the organization are improving.

Part 2 is concluded with a chapter on contributing to the community wide pursuit of excellence. (Chapters 4–9).

Part 3, "Quality System Desk Reference," outlines the elements of a quality management system for controlling and improving quality, supporting the Total Quality culture, and assuring customer enthusiasm. This desk reference is included to provide managers with an overview in sufficient detail that it can be used to assist them in assessing the adequacy of their overall management system. It consists of 13 sections.

Part 4, "Appendices—Examples," provides a discussion of several powerful yet simple problem solving techniques, and examples of annual strategic plans and various quality system procedures and processes.

Important Considerations

It is customary in western society to denigrate the focus on short-term gains and espouse the need for a long-term outlook. However, when based on the quality precepts addressed in Chapter 2, both of these become important parts of moving forcefully into the future. Lets clarify this clarification.

The quality transformation process leads to an organization with the long-term outlook. Implementing this process carries with it the need for an up-front investment. When combined with a parallel focus on short-term gain through quality improvement (waste reduction), an organization is able to quickly generate cost savings which are used to substantially fund the transformation. If this waste reduction effort is based on the quality precepts and follows the precautions addressed in Chapter 5, it will also help prepare the organization for the cultural change. In other words, it will help pay for and prime the pump.

This book addresses the transformation process, and not quality-related short-term gains. Techniques for identifying and reducing waste are addressed in the book but will primarily be learned as a result of implementing the investigative phase in Chapter 4. This knowledge should be selectively applied to achievement of short-term quality gains.

This book is intended to be applied in a graded fashion based on the size of the organization, the importance of the specific products/ services and system elements, and its impact on the customer, organization, and society. Although certain portions of this book may not appear to be appropriate in your circumstances, careful consideration should be given before excluding any element (understanding the tools and philosophy will generally lead to the recognition that it is applicable).

The very existence of a document providing generic guidelines for managing quality can imply incorrectly that one basic management system will suffice in a given industry or situation. A management system developed for a specific application should be used with extreme care when applied elsewhere in that industry or when used to evaluate the adequacy of another organization's management system (or better, strictly avoid such applications).

There exist countless specialized disciplines related to elements identified in this book and other important management and technical functions. Examples of such areas include reliability, configuration control, human resource management, accounting, and statistical quality control. Some of these areas may be mentioned in this document. However, the reader should refer to other documents or professionals for detailed guidance on the disciplines.

Special Word Usages

In keeping with the intent of this book, the use of industry specific terminology has been minimized. Where questions arise related to the meaning of specific words, a standard collegiate dictionary should provide the needed guidance. The following word usage clarifications are provided to assist in interpreting this book by amplifying specific aspects of the definitions which are important in this book.

Organization

A company, corporation, firm, or enterprise, whether incorporated or not, public or private. It may also apply to any segment of a large organization.

Customer

The consumer, user, client, or beneficiary of the product or service. A customer may be external to, or within the organization. Internal customers are the people in the organization who use or build upon an individual's work (people further along in the production process, such as those who use data you generate or read your reports).

Supplier

An organization that supplies product or service. In general, this term is used to designate an organization which supplies product or service to your organization.

Service

The interaction between a supplier and a customer which is meant to satisfy a stated or implied customer need. Generally speaking, a service is an intangible provided to a customer; however, delivery of tangible goods may form part of the service and the supplier may be represented by equipment during the interaction (such as a vending machine). Likewise, service is a part of the production and sale of tangible goods (for example: marketing, billing, shipping, field support, and customer complaint handling).

Product

Raw and processed materials, parts, and equipment; tangible goods.

Quality

The totality of features and characteristics of a product or service that bears on its ability to satisfy stated or implied needs. It encompasses safety, performance, dependability, timeliness, value, and productivity of the product or service and the activities associated with production of the product or service.

Total Quality

A way of doing business—an all-encompassing quality-focused approach which creates and gains its advantage from a synergy among all aspects of the organization working together to achieve excellence. An approach which creates value for customers, employees, stockholders/ owners, and the community, and which ultimately leads to a realization that products and services are the expression of human excellence.

Quality Policy

The overall intentions and direction of an organization as regards quality. This term should be generally understood to apply to only those policies which are formally expressed.

Quality System

The organizational structure, responsibilities, procedures, processes, and resources for managing quality. It provides the basis and structure for the overall management system.

Production

The act of producing product or service for internal or external use, including administrative activities.

Team Facilitator

The team facilitator is a nonteam member who

- Provides training to the team leader in team dynamics and team leadership; provides training to the team in team process and techniques; and
- Acts as an independent observer of team activities, as a coach to the team leader (including counseling on team activities and performance), and as mentor to the team (including intervening in and diffusing destructive team situations).

The ultimate objective of the team facilitator is for the team to evolve into a self-facilitated team; at which time the services of the facilitator will no longer, or rarely, be needed.

PART 1

Quality Precepts and Quality System Principles

Before starting the path toward excellence, it is important to be familiar with the principles upon which the path is founded and the structure which lies along the path.

This portion of the book provides information on the general precepts of quality and quality management. It addresses the responsibilities of management and employees in the production, control, and improvement of quality. It also addresses principles of the quality management system, discusses the relationship between the overall management system and the quality management system, and provides an outline of the system components.

CHAPTER 2

Quality Precepts

In this chapter, you will find a list of ideals which provide the basis for achieving excellence. When practiced, promoted, and supported consistently by top management, these ideals will become the organization's style and culture.

Consider the preceding sentence again. It addresses two very important concepts.

- The pursuit of excellence requires a cultural shift in the organization, a change in attitudes regarding quality. Cultural change takes time and is resisted by many people. To ensure the success of a quality system, it is crucial that management control the timing of system changes so as not to out-pace the cultural changes. This topic will be addressed throughout the second part of the book.
- Cultural change must start at the top of the organization. Leadership is the key. Upper managers must demonstrate (embody) the concepts before expecting their people to accept the concepts. Be-

fore the cultural change is underway, management must be consistent in word and action.

As a prelude to this chapter, consider President Ronald Reagan's statement at the first Malcolm Baldrige National Quality Award Ceremony:

> The one trait that characterizes these winners is that they realize that quality improvement is a never-ending process, a company-wide effort in which every worker plays a critical part . . . These awards are won by companies, but they are earned by individuals working together in the quest for excellence.

Note: The Malcolm Baldrige National Quality Award was established by an Act of Congress to promote general awareness of the importance of quality in the global marketplace and to recognize those companies that practice Total Quality management strategies.

Internal Quality

The source of quality lies within each person. The pursuit of excellence starts as an inner process, with the impulse for creative self-expression; it manifests itself as the tangible expression of human excellence. The pursuit of excellence by an organization must be based on and built around the pursuit of human (individual) excellence.

Customer Focus

The ultimate focus and objective of everyone in the organization must be on satisfying the customer. This applies to internal as well as external customers. Part of accomplishing this focus is ensuring that everyone understands how their activities affect the product or service and hence the customer.

Quality Context

Teamwork, innovation, accomplishment, and integrity form the basis for relationships within the organization.

Leadership is the key to excellence. The aim of management must be to help people to perform and improve their job. Leaders focus on im-

also lead people to withhold data which reflects poor performance. Feedback should also reflect an understanding of the situation (in general, this requires historical data and the application of statistical methods).

Motivation and dedication are supported through encouraging people, acknowledging efforts, and providing equitable rewards for significant accomplishments. Encouragement is possibly the most neglected and cheapest of the three; people should be encouraged to pay attention to quality within their area and to use their ingenuity to improve the quality of their products, services, and activities. (On a regular basis, let people know in little ways that their efforts are appreciated. For example, hand out cards with a thank you and quality-focusing thoughts when someone is seen giving that extra effort, or anonymously give small gifts in recognition of special efforts.)

Communication

Communication is a two-way process. It includes listening and a willingness to hear the unvarnished truth (or opinion, right or wrong) without criticism. It needs to be based upon basic individual equality and worth, and an honest desire to understand and help.

It is essential that people are allowed and encouraged to contribute their ideas about goals and ways to achieve the goals, and are given the responsibility and authority for achieving the goals. To a large extent, this is accomplished by including all personnel in the planning process.

Mechanisms must be established for continuous vertical and lateral communication within the organization, communication with customers, and communication with suppliers. One mechanism which is effective in each of these areas is the use of improvement and project teams. Another is the prompt issuance of bulletins to inform people who may be even remotely impacted by a decision or event (the key here is complete honesty), to provide recognition, and to keep everyone generally informed regarding the functions of the organization (this is also an excellent rumor control mechanism).

Barrier Elimination

Obstacles to excellence include focus on the short term, resistance to change, inconsistent management signals, an unbalanced focus on self-interest, us-them attitude, and fear.

Emphasis is placed on prevention of problems. Firefighting directs attention to quick fixes which seldom eliminate the real cause for the problem; hence firefighting is self-perpetuating and results in a massive waste of resources, profit, and morale.

Management must be willing to make up-front investments for long-term quality improvement and accept the cost of managing quality as a normal cost of business. Perhaps the biggest up-front cost when starting cultural change will be the cost of educating management from the top down. Although operational costs generally rise in the beginning of the quality transformation, the true financial impact of ongoing quality improvment becomes dramatically clear within a few years. A smaller percentage of resources will be used in locating new customers because fewer will be lost and the customers will become the primary source of new customers. There will be a significant reduction in wasted resources because the production, administration, and management processes will be more efficient and effective. There will be fewer firefights, and more of the product and services will be error free. Creativity will be unleashed and harnessed for the creation of new, better, and more profitable products, services, and markets.

Management must maintain a consistent focus on, and support of, quality control and improvement. Inconsistency sends the signal that quality focus is only a management fad which will soon disappear; and it leads to token support in the organization, at best. Commitment and consistency in these areas sends the signal that the organization's culture is changing, and accelerates the momentum for change.

Rewards and advancements must be provided to those individuals and groups who consistently place success of the whole at the top of their list of self interests. Concern for the well being of society as a whole, the community, the environment, and the customer should take precedent, followed closely by concern for the well being of the organization, fellow workers, suppliers, and self. True success can only be achieved through an awareness and understanding of the interrelationship with, and interdependence on each of these. (The theory of suboptimization states that the whole can be optimized only when some of the parts are suboptimized. Consider the effect on your organization and the customers should purchasing base it's decisions on price alone.)

People cannot work effectively when they fear that their basic needs are in jeopardy. Everyone needs to realize and expect that improvement leads to greater security, not loss of jobs. Everyone must feel and demonstrate that acknowledgment of mistakes leads to improvement, not pun-

ishment. Fear is reduced through consistent appropriate management intent, action, and open communication.

It is important that team participation be encouraged. Single mindedness should be combated through leadership (setting the proper example and private counseling). Single mindedness should rarely be punished— the focus must be on encouraging growth. (There are people who work best as individuals. This should be recognized and allowed where appropriate.) It is also important to remember that the strength of the whole results from a combining of the strengths of the individuals involved.

Improvement

Improvement is an ongoing effort which focuses on all aspects of the organization including financial and human resource management; administration, production, and management processes; products and services; relationships with customers and suppliers; and relationships between departments and between people.

New knowledge is at the heart of continuous improvement. New knowledge is obtained through continued learning which includes the collection and analysis/evaluation of information. Activities, methods, and tools which enhance the rate at which new knowledge is generated (enhance the rate of learning) and applied are the keys to improvement.

Continuous improvement has a compounding effect on quality as one improvement is built upon another. Maximum benefit is gained when the management system fosters spontaneous innovation and provides a structured process for planned improvement. (Communication of improvement goals and progress throughout the organization and with other parties is an effective mechanism for increasing involvement in and support for continuous improvement.)

Problems are opportunities for improvement. By focusing on the positive, it opens the mind to see the cause of the problem, thereby allowing actions to be taken which reduce the likelihood of recurrence. Such an approach also overcomes a tendency to focus on the flames (and blames).

No two items are perfectly identical. Variation is a fact of nature. However, variability can (and must) be reduced in the organizational setting. A major aspect of improvement is the reduction in the variability of: the operation and performance of all processes, the resulting products and services, purchased products and services, customer satisfaction, etc.

Feedback from customers of their experience of product and service quality, as well as from monitoring of internal processes, is essential in evaluating variability and ensuring customer satisfaction.

CHAPTER 3

Quality System Principles

The pursuit of excellence by an organization requires an integrated management system (involving all elements of the organization) based on quality precepts. Quality management is an integral part of this system with the top manager being the custodian of, and the entire organization being perceived as owner of, the quality management process. It means doing things such as:

- Providing leadership from the top
- Establishing vision and policies
- Establishing improvement goals
- Establishing performance measurements
- Providing feedback on performance
- Providing resources and training
- Providing rewards and recognition
- Establishing requisite control over production, administration, and management processes

Note: The use of improvement goals should not be confused with "management by objectives" as it is customarily applied. Improvement goals focus on changing the causes such as achieving statistical control of outstanding debts, reducing variation in particular activities and processes, and enhancing the capability of people and machines. In short, the focus is on *how* the goal is to be achieved.

Control versus Improvement

An effective management system has two basic aspects. One focused on control and the other on improvement. One without the other will inevitably lead to failure. Both are required to meet customer needs and expectations, and ensure long-term viability of the organization.

The control aspects of the system ensure that we stay on course, activities are performed in a uniform, consistent manner, regulatory and contractual requirements are fulfilled, and lessons learned do not need to be relearned. Control implies the existence of an established norm. It provides assurance that life will be predictable and free from unpleasant surprises.

The improvement aspects of the system ensure long-term growth and security of the organization through productivity enhancement, product development, product and process improvement and innovation, and market expansion. Improvement means change to higher levels of performance. Some call it breakthrough because it reflects a decisive, dynamic movement to new, higher levels.

Another way of understanding the difference is that control is the prevention of bad changes and improvement is the creation of good changes.

Control and improvement form a continuing cycle of events consisting of alternating plateaus and performance gains.

Relationship with Overall Management System

To start with, lets review the definitions of quality and quality system. Quality is:

The totality of features and characteristics of a product or service that bears on its ability to satisfy stated or implied needs. It encompasses safety, performance, dependability, timeliness, value, and productivity of the product or service and the activities associated with production of the product or service.

All activities, products, services, and processes have characteristics by which quality is judged, including those associated with financial management, marketing, and human resource management. In addition, every function within an organization provides products or services either directly to the outside customer or in support of those functions which do. As such, the quality system must be applied throughout the organization.

The Quality system is:

The organizational structure, responsibilities, procedures, processes, and resources for managing quality. It provides the basis and structure for the overall management system.

Although the first sentence in the above definition is relatively clear, the second sentence deserves further explanation. It underscores the importance of thoroughly integrating the quality system into the overall management system and of ensuring that the entire organization is perceived as owner of the quality management processes.

Next let us consider the foundation for the overall management system, which includes:

• An organizational definition and establishment of responsibilities which divide the organization into functional units, and
• Policies and procedures for directing and controlling the functions of the organization.

From the definitions, we see that the overall management system and the quality system are, in effect, one and the same. It is therefore important that all members of the organization understand that the quality (or management) system is owned by the entire organization; although individual elements of the overall system are owned by those functions or individuals which have primary responsibility for their implementation (such as the marketing manager who owns the elements which control the marketing functions).

Quality System Overview

Even though an effective quality system may be perceived as being holistic, development of the system depends on focused attention to each part and its relationship to the whole.

Numerous books and standards elaborate on the elements (or elemental parts) which should be addressed in a quality system. In general, each element generically addresses a specific functional area of quality management and a complete set of elements theoretically addresses all areas which are important in managing quality within a business.

Each organization combines these elemental parts in a different way to form a set of quality programs which define its overall management system. This set is generally referred to as the "Quality Program." Once documented, each program provides upper level policies, requirements, and guidance; it also provides a general description of the implementing mechanisms or may provide the implementation detail for use throughout the organization. Each program may or may not relate directly to an element. In fact, many of the elements are best handled by combining them into one quality program while others may be integrated into several quality programs. For example, all elements related to procurement are frequently combined into a Procurement Control Program or Procurement Quality Assurance Program. These programs are implemented as applicable by all segments of the organization, either directly from the program document or through procedures which contain implementation details.

Consider a program for the control and corrective action of problems/concerns. This program could be comprised of system elements dealing with internal problems/concerns and those dealing with customer concerns (see Sections 11 and 13 in the "Quality System Desk Reference," on pages 118 and 124, respectively). The documented program could contain the policies, requirements, and general description of the implementation mechanisms and interfaces; in such a case, one or more procedures would be issued to implement the program. As an alternative, the documented program could be written to be implemented directly from the program document as in the example in Appendix V (page 171).

The third part of this book, "Quality System Desk Reference," groups the basic elements of a quality system into general categories as an aid to understanding the individual functions and interdependence of the parts. For example, the quality system may be seen as performing three primary functions:

1. Providing mechanisms which focus all members of the organization on the same objectives. Examples include the organization's mission statement and the annual strategic plan. These mechanisms are outlined in Section 2 of the "Quality System Desk Reference" (page 88).
2. Providing mechanisms which ensure efficient and controlled pursuit of the objectives. Examples include document control, market analysis, process improvement, and structured teams. These mechanisms are outlined in Sections 4–13 of the "Quality System Desk Reference" (page 95).
3. Providing mechanisms for assessing progress and effectiveness (or quality assurance). Examples include quality audits and performance monitoring reports (or executive summaries). These mechanisms are outlined in Section 3 of the "Quality System Desk Reference" (page 91).

While each of these primary functions (categories) are predominantly addressed in different sections of the desk reference as indicated above, the interdependence is addressed throughout this book; and it is actually the mechanisms supporting focus and assessment which are integrated into the *quality ensurance* category that accounts for the apparent overdominance of the quality ensurance category in this book.

Before reading further, you may find it helpful to become more familiar with the elements of the quality system. This may be accomplished by briefly reviewing the "Quality System Desk Reference" (page 82). During this review, focus on obtaining a general understanding of the element groupings and becoming aware of their interrelationships.

PART 2

Pathway to Excellence (Quality Culture and System Development Process)

Now that we are familiar with the principles upon which this path is based, we are in a position to understand and therefore follow the path. So let's get started.

This portion of the book describes the process for nurturing a Total Quality culture, and developing and implementing a quality system which is geared to achieving excellence in all aspects of the organization (Total Quality). It also references examples which are contained in the Appendices.

Once again, it is important to remember that an effective quality system is built and evolved over time,

- As management grows in their understanding of Total Quality, and
- In step with the evolving attitudes of the organization's people, customers, and suppliers.

An effective quality system is built and evolved using a process similar to building a house:

1. Evaluate needs, investigate alternatives, and plan,
2. Establish the foundation and revise/expand the plan,
3. Build the structure and revise/expand the plan,
4. Complete the house,
5. Live with, maintain, and improve the house.

The same process is used for rebuilding when the house is to be placed in the showplace of homes.

Important Considerations on Implementing This Process

If your organization is small, you may wonder how you can possibly follow this path (implement the transformation process). Actually, you may find it easier than would a large organization because your final management system will be simpler and you will have fewer personality constraints to contend with.

Notes have been placed throughout this part of the book to help you understand how to scale down the process. These notes are identified by the prefix SO (for Small Organization). Many of these notes will also be useful to organizations of all sizes as every organization will scale down some parts of the process and systems.

Most of the examples were taken from a small sole proprietorship (the Communication Clinic, Chattanooga, Tennessee) which thinks big and started early preparing for the inevitable growth. The company started as an individual speech-language pathologist with an interest and experience in public speaking and interpersonal communication. Within six years the company had seven speech-language pathologists, had expanded its scope to include seminars in effective communication, and had received national recognition for it's efforts in training individuals and groups to be effective communicators. It was at this stage that the company initiated development of a management system based on the pursuit of excellence.

To ensure consistency in the examples and thereby facilitate understanding of the relationship between examples, the examples have been largely revised to minimize any distinction between those examples developed specifically for this book, those provided by the Communication Clinic, or those provided by other organizations.

References to sections in the "Quality System Desk Reference" are identified by brackets [].

PATHWAY TO EXCELLENCE

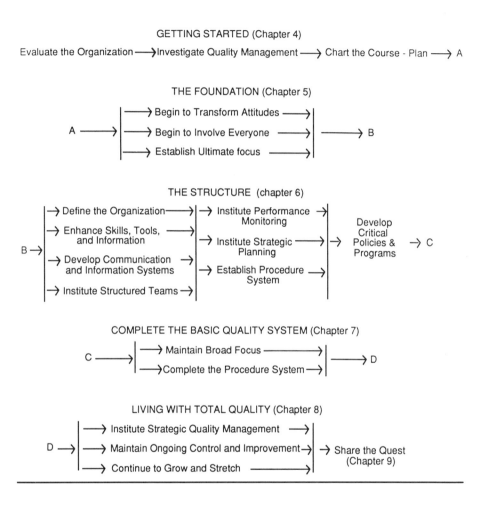

GETTING STARTED (Chapter 4)

Evaluate the Organization ——⟩ Investigate Quality Management ——⟩ Chart the Course - Plan ——⟩ A

THE FOUNDATION (Chapter 5)

A ——⟩ | ——⟩ Begin to Transform Attitudes ——⟩ |
       | ——⟩ Begin to Involve Everyone ——⟩ | ——⟩ B
       | ——⟩ Establish Ultimate focus ——⟩ |

THE STRUCTURE (chapter 6)

B ⟶ | ⟶ Define the Organization ——⟩ | ⟶ Institute Performance ⟶ |
     | ⟶ Enhance Skills, Tools, ——⟩ | Monitoring | Develop
     | and Information | ⟶ Institute Strategic ——⟩ | ⟶ Critical   ⟶ C
     | ⟶ Develop Communication ——⟩ | Planning | Policies &
     | and Information Systems | ⟶ Establish Procedure ——⟩ | Programs
     | ⟶ Institute Structured Teams ⟶ | System |

COMPLETE THE BASIC QUALITY SYSTEM (Chapter 7)

C ——⟩ | ——⟩ Maintain Broad Focus ——⟩ |
      | ——⟩ Complete the Procedure System ——⟩ | ——⟩ D

LIVING WITH TOTAL QUALITY (Chapter 8)

D ——⟩ | ——⟩ Institute Strategic Quality Management ——⟩ |
      | ——⟩ Maintain Ongoing Control and Improvement ⟶ | ⟶ Share the Quest
      | ——⟩ Continue to Grow and Stretch ——⟩ | (Chapter 9)

CHAPTER 4

Getting Started—Evaluate, Investigate, and Plan

Evaluate

A small team should be formed, preferably one which includes the top manager. Take an honest look at the organization to identify its strengths and weaknesses. Avoid getting bogged down in the details and in the search for solutions; this evaluation should be aimed at obtaining an overview of the strengths and weaknesses in preparation for outlining the initial plan.

Compare the management system and style, work environment, and management, employee, and customer attitudes to the information provided in Chapters 2 and 3 and in the "Quality System Desk Reference." For example,

- Are the organization's mission, unique competency, and long-term objectives clear? Do they reflect the quality precepts? Are they documented for all to know? Is everyone associated with the organiza-

tion (including customers and suppliers) aware of this information? Is it clear that the people in the organization understand and are actually pursuing these objectives?

- Is customer satisfaction the preeminent consideration?
- Do the people in the organization consider others within the organization as their customers and demonstrate this awareness through their actions?
- Do teamwork, innovation, accomplishment, and integrity form the basis for relationships, both within the organization and with customers and suppliers?
- Is management style based on leadership or does it reflect another form of management such as benevolent dictatorship? Does everyone in the organization participate in organizational goal setting and planning, and are they involved in establishing policies?
- Does the organization fight fires or focus on preventing problems?
- Is the organization investing in the future or driving up short-term gains?
- Is there a mechanism for problem/concern documentation, evaluation, resolution, and recurrence control? Is management really informed as to the nature and scope of problems and customer concerns?
- Is there an effective systematic mechanism in place to ensure that everyone knows what is planned or just happened, or is gossip the predominant mechanism?
- Is there an established procedure system? Do people comply with the procedural requirements and guidance? Is there a mechanism for assessing compliance with and effectiveness of the procedures?

Note: If top management is committed to pursuing Total Quality from the start, the evaluation (above) and investigation (below) phases of the getting started process may be combined and performed at the same time; in which case, startup time may be substantially reduced.

Investigate

The team should now be expanded to include all top management with the top manager functioning as team leader. The team may also include other selected people from the organization who are recognized as leaders and achievers, fill critical positions, or are otherwise highly influential (such as union representatives). If the top manager is not an effective team leader and is willing to accept guidance from another person, the team may be facilitated (see "Special Word Usages" in Chapter 1) by a nonteam member who is knowledgeable and experienced in team dynamics and team leadership. This team should be chartered with the ongoing responsibility for quality management within the organization (henceforth, this team will be referred to as the Quality Council).

One member of the team should be assigned responsibility for coordinating the Total Quality effort, and this responsibility should be clearly established as his/her top priority. In large organizations, it is generally better for this person to be assigned to a position dedicated to fulfilling this responsibility (such as Total Quality Coordinator). The functions and duties of the Quality Council may also be supported by technical, administrative, and clerical personnel who directly report to the council or the Total Quality Coordinator on an as needed or permanent basis.

SO Note: The Quality Council should have at least two people on it. If there is only one manager in the organization, include a nonmanager who is a leader among the other employees and who is an achiever.

The investigation phase of getting started is essentially a self-guided course on quality management. Course details such as syllabus, reading list, and the various quality techniques and guru approaches to be studied are the responsibility of the Quality Council. Detailed suggestions are not provided in this book. Use of this truly self-guided course approach enhances the use of creative self-expression, minimizes the potential for imposition of "outside" biases, and promotes personal experience, practice, failures and new tries (the path to true understanding), and ultimate success.

One obstacle in gaining true understanding is a general tendency for people to listen for things they already "know" to be true. As such, learning is frequently biased by preconceived notions or "conventional wisdom." Members of the Quality Council must be aware of this natural bias and focus on approaching every idea as if it is new and personally unexplored. This approach is essential to achieving a cultural transformation. (It also leads to the development of a management characteristic which is important to propagating the Total Quality culture, in particular, a natural "open-minded" response to people and ideas.)

The following should be defined as the initial objectives of the Quality Council.

1. To become well versed in quality precepts and the principles of quality management. This may be accomplished by the following; however, it is important that the council experience a variety of viewpoints (avoid getting hooked on one "guru"):
   • Review and discuss Chapters 1–3 of this book,
   • Develop a reading list which each member of the council is to read (the Bibliography at the end of this book may serve as a starting point), and discuss the topics and ideas,
   • Attend a variety of seminars geared to upper management.
2. To become familiar with the transformation process. This may be accomplished by review and discussion of Chapters 4–8.
3. To obtain training on team leadership and team problem-solving process and techniques.
4. To become familiar with a variety of tools and techniques that are used in managing quality, with basic statistical methods being among the most important.
5. To become familiar with the functions and mechanisms of quality systems. This may be accomplished by the following, after the council has made some headway on objective 1:
   • Review and discuss the information presented in the "Quality System Desk Reference,"
   • Expand the reading list,
   • Attend seminars geared to quality auditors and managers.
6. To become familiar with the scope of applicable laws, regulations, and industry standards which could have an impact on, or provide guidance for, the organization's products, services, and activities.

Note: Completing these objectives may require more time than you think is necessary. If this is what your are thinking, RECONSIDER. This is a critical part of the transformation process. Pursuit of these objectives facilitates cultural change among the members of the Quality Council, and provides the knowledge which will overcome the handicap caused by inexperience in any specific area. The time invested here will be more than made up during the rest of the development process and will greatly enhance the likelihood that your organization will complete the process.

Once the Quality Council has fulfilled these objectives, the council members should continue to pursue the objectives and share their new discoveries with the other members of the council. Possibly the best method for accomplishing this ongoing pursuit is to join organizations focused on quality and productivity, and to subscribe to quality focused magazines. One such organization is the American Society for Quality Control based in Milwaukee, Wisconsin, with local chapters throughout the United States. Members in this organization receive a monthly magazine, *Quality Progress*, which is an excellent source of ideas, philosophy, and experience; the members are also made aware of meetings, conferences, and seminars for the quality professional where ideas can be exchanged and detailed instruction obtained. Other such organizations include the American Productivity and Quality Center located in Houston, Texas, and the Association for Quality and Participation located in Cincinnati, Ohio. *Note:* Your local chapter of the American Society for Quality Control may also be consulted for help in developing reading lists and locating seminars.

Plan—Charting the Course

As the Quality Council gains greater understanding of quality management, the council should start to outline a plan for transforming the organization. Major aspects of this plan are addressed in this chapter and Chapters 5–8.

It is important to understand that this will be a living plan, in that it will grow and evolve as greater portions of the organization begin to participate and add their ideas and insights.

Primary focus now should be placed on strategy for establishing the foundation and structure (discussed in Chapters 5 and 6):

PLAN AHEAD

- Changing attitudes—incorporating quality precepts into attitudes and actions,
- Involving all members of the organization in the transformation process,
- Defining the organization's mission, distinctive competency, and long-term objectives,
- Defining the organization,
- Developing programs to enhance the rate of learning,
- Developing structured team programs and refining communication and monitoring systems,
- Outlining the procedure system and developing critical policies and programs.

In developing this strategy, consideration is given to factors which may inhibit and promote the transformation. During your training on team problem solving, techniques will be learned for evaluating inhibiting and promoting factors, maximizing the promoting factors, and minimizing or converting the inhibiting factors. The Pareto principle, brainstorming, force field analysis, why-why diagrams, how-how diagrams, and fish bone Analysis are some of the more popular problem solving techniques. (These techniques are discussed in Appendix I and are addressd in many books on employee involvement, team processes, and Quality Circles. Visit your library, contact your local chapter of the American Society for Quality Control, or contact the Association for Quality and Participation in Cincinnati, Ohio.)

The plan should also provide an overall outline for the transformation process and initial estimates for the resources and schedule. Let's emphasize the word *initial.* While providing general targets, the plan must be expected to change to reflect experience and changing situations. The only restriction is that at no time should the organization's members be given the impression of fluctuation in management's commitment to quality and the transformation.

The following are *rough* estimates for various segments of the transformation process. The word rough is emphasized because only the Quality Council has the critical information; such as, extent of change needed in attitudes, extent of structural changes, educational needs within the organization, size and layering of the organization, resources (human, mechanical, financial, etc.).

1. Evaluating, investigating, and planning .... 1 year
2. Establishing the foundation and structure, including initiation of
   • Pilot problem-solving/improvement teams
   • Communication and monitoring systems development
   • Programs focused on developing/obtaining and using new skill/tools .... 1 year
3. Expanding the use of problem solving/improvement teams throughout the organization .... 0–2 years
   SO Note: Small organizations may only have one team, which is also the pilot team.
4. Issuing new/revised procedures which define and implement the management system .... 1–3 years
5. Completing documentation of the management system, and refining (debugging) the annual strategic planning process .... 2 years

The above timing estimates indicate the general sequence of events leading up to achievement of a Total Quality culture and effective quality management. However, due to overlapping and with effort, the entire sequence may be completed in four to five years with significant payoffs being realized in the third year.

Note: If your Quality Council decided against the time investment discussed in the previous section on investigation, then the entire sequence may take twice as long (or may never be completed).

CHAPTER 5

Establish the Foundation

Transforming Attitudes

The attitudes of the organization's members are the primary determinants in the future of the organization, including failure, boundless growth, and every possibility in between. Management's failure to recognize the importance of attitude and to foster a change in the attitudes is the primary reason for the failure of the quality transformation process. Failure can also occur when implementation of the quality management structure outpaces acceptance of the change and understanding of the need and benefits.

Attitude transformation should start quietly and slowly with the Quality Council leading the way. Slogans and exhortations often (almost always) do more harm than good. The members of the Quality Council should start to live and work in accord with the quality principles. They need to *consistently:*

- Exemplify a concern for customer satisfaction, both external and internal,
- Exemplify care for the organization's people,
- Maintain focus on quality,
- Demonstrate and encourage teamwork, innovation, and process improvement,
- Place greater responsibility and authority in immediate reportees, but be sure their current capabilities are not overly extended (stretch, not break),
- Demonstrate and focus attention on prevention whenever fires occur, and exercise awareness and restraint so as to not initiate firefights by others,
- Focus attention on resolving problems, not identifying blame,
- Praise in public, counsel in private (even in praise, apply caution. People from some cultures find public praise extremely embarrassing and may avoid actions which lead to public praise.),
- Discreetly ensure that the people with whom they interface are aware of recent participation in quality training,
- Encourage everyone to participate in continuing education programs, support their efforts, and arrange for in-house miniseminars with open enrollment (please notice this *did not* say required attendance),
- On a regular basis, reach out in little ways to let people know their efforts are appreciated,
- Communicate, communicate, communicate.

As sincere interest in quality grows, the Quality Council should respond to the new needs of the organization's people by providing generalized training in quality precepts, and training in the fields of management, process control, improvement, statistics, and the technical aspects of the organization. Personnel should also be given time to share and contemplate their new knowledge, and be encouraged to use their knowledge (use it or loose it). Once again, people should be encouraged to participate, not forced. These actions will not only increase the pace of attitude transformation, but will also encourage innovation, teamwork, and the focus on improvement. (Training is discussed further in Chapter 6, see "Skills, Tools, and Information" and "Structured Teams.")

After the majority of those in the organization are well on their way to the new set of attitudes, discreetly suggest that *they* develop a motto for the organization which reflects their new approach to work life.

Then support their effort, possibly even providing a celebration (prize for everyone) to proclaim the motto. Above all, avoid giving the impression that management is directing the effort. One way to accomplish this is to ensure that management is in the minority on the motto team, and encourage the team to narrow down the choices and let the winning motto be selected by organization-wide secret ballot. Internal competition should also be avoided; the focus must be on developing cooperation and a family spirit.

SO Note: It is generally better for very small organizations to use consensus in selecting the motto.

Note: The principle difference between a motto and a slogan/exhortation is that the motto reflects pride in the current culture and organization, whereas, a slogan/exhortation is used by a small group of people as a tool in driving change. Again, slogans and exhortations should be avoided.

Involvement

Involvement is the foundation for growth and enhances employee satisfaction (quality of work life). Everyone in the organization needs to have and feel that they have an involvement and influence on the quality of products and services provided to customers. They need to feel a sense of responsibility for, and ownership of, the organization.

Imagine the cost reductions which will occur when everyone takes responsibility for the success of the organization, when everyone is improving the effectiveness of their functions, when people work together to improve their functional interfaces and common processes. Imagine the increase in product and service quality and market expansion which will occur when people are routinely stimulating and building upon each other's creativity.

The number of ways which involvement can be accomplished is only limited by the creativity of the Quality Council and the other people in the organization. Involvement may be accomplished through:

- Participation in development of ultimate focus [Subsection 2.1, page 88], policies [Subsection 2.4, page 89], and strategic planning [Subsection 2.5, page 90],
- Participation in establishing personal work objectives, schedules, and methods,

- Multidisciplinary project teams,
- Quality improvement teams,
- Quality circles,
- Cross-functional process teams,
- Problem response and recurrence control teams,
- Customer and supplier interface teams,
- Work unit participative management teams.

The Quality Council should always be searching for ways to enhance involvement. It should periodically assess and, as appropriate, modify involvement mechanisms.

There is one very important inhibiting (potentially debilitating) factor which the Quality Council should consider. When people feel that involvement leads to management gain and employee loss, morale drops and their involvement will probably result in a waste of resources at best, and destruction of the organization at worst. There are many ways to prevent or substantially reduce the possibility that this will occur. For example:

- Establish enhancement of employee satisfaction (quality of work life) as a primary objective of employee involvement.
- Ensure that involvement is voluntary (involvement will automatically improve as attitudes improve), *Note:* Beware, sometimes personal zeal leads to a tendency to overencourage involvement or insinuate that noninvolvement will be frowned upon. Such actions can lead to widespread belief that involvement is functionally nonvoluntary.
- Allow teams to select their own projects (or when a team must be formed to address a specific problem or project, focus on voluntary involvement),
- Limit the participation of supervisors in teams which substantially involve people below them in the organizational hierarchy (the importance of this diminishes as people become accustomed to team activities and supervisors more consistently demonstrate leadership),
- Provide incentives linked to the success of the organization, *Note:* Beware, incentives linked to the success of organizational units or functions tend to stimulate individual success at the cost of the whole organization.
- Consistently demonstrate through word and action that improvements in effectiveness will lead to greater employment security, career development, and monetary rewards, not staffing reductions (and mean it).

Like attitude transformation, involvement should be started quietly and slowly. The members of the Quality Council should start to use involvement techniques during the "Plan—Charting the Course" phase (Chapter 4). They must set the example (discreetly at first, then with increasing consistency) by:

- Working with their peers on projects which cross functional bounds,
- Working with people from their own organizational unit and other affected units on important projects as a team, which includes decision by consensus, *Note:* Beware, it is important that the managers function as leaders as opposed to bosses. As such, it is helpful if managers have had some team leadership training before participating on teams with those below them in the organizational hierarchy. Also, managers should be cautious that they do not usurp the

perceived or actual authority of others in the organization when they initiate the team.

- For specific projects over which their immediate reportees have authority, recommend that the projects be closely coordinated with the other people who are involved or may be impacted by the project, and, as an aside, suggest that project be approached as a team (this may need some explanation). *Note:* Avoid becoming discouraged, people must learn to work together as equals through visible example, training, and practice.

The next step is to prepare for the use of structured teams. This is accomplished by providing team leadership training to several people who are natural leaders and achievers, and are willing to participate in the training. Once trained, these people should become more and more involved in the discrete team activities which are mentioned above. These people should be given experience as both team members and leaders. ("Structured Teams" is addressed in Chapter 6.)

Another major part of ensuring involvement is structured participative management. It is generally accomplished by building participation mechanisms into the management systems such as defining the organization; developing goals, policies, and procedures; improving processes; and career planning. Structured participative management is initiated below in "Establishing Ultimate Focus," and will be enhanced in Chapters 6–8.

Establishing Ultimate Focus [Subsection 2.3, page 89]

Organizational success is strongly dependent upon the extent to which its efforts and resources are concentrated or diffused. For example, imagine the cost and quality of your home if each member of the construction crew received a different version of the house plan.

Every organization encompasses many different functions which must be coordinated and headed on a common *mission* in order to ensure that the mission is accomplished in the most expedient and rewarding manner. Consider this mission as a point in space.

To focus efforts and resources, some additional guidance must be provided to ensure that all resources and efforts are approaching the mission from a similar direction. This guidance should be provided by

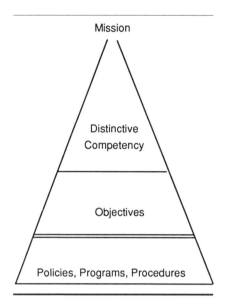

defining the *distinctive competency* of the organization, or that which sets it apart from other organizations which may be on the same mission. The mission now forms the pinnacle of a cone.

To ensure that the organization maintains its focus within this cone, *objectives* should be established which relate to, and provide guidance for, all aspects of the organization. The objective statements now form the base of a cone which defines the ultimate focus of the organization. This cone will be extended by the policies, programs, annual strategic plans, and procedures which are addressed in Chapters 6–8.

The objective statements need to

- Define the ideal focus for the organization,
- Reflect the work ethics, morals, and aspirations of the organization's members, and
- Be supported by the organization's members.

To accomplish this, the Quality Council should inform the members of the organization of the nature and purpose of the statements, encourage them to provide ideas, and encourage them to review and comment on the draft statements.

Note: At this point, the Quality Council has given considerable thought to quality and probably experienced considerable change in quality precepts and attitudes; while the remainder of the organization's members have yet to experience the same opportunity. As such, the initial ideas which are provided to the Quality Council should be interpreted by the council based on their enhanced understanding. In a sense, the council will *discreetly* lead the other members of the organization to the ideal focus which will reflect their future work ethics, morals, and aspirations.

The following are examples of a mission statement, statement of distinctive competency, and objectives.

Mission Statement

Communication Clinic exists to educate people in the importance of effective communication and help them improve their ability to be clearly and readily understood.

Statement of Distinctive Competency

Communication Clinic pursues all aspects of communication, focuses on the whole person, and believes everyone has the ability to improve.

Objectives

1. To continually and enthusiastically pursue innovation, improvement, and excellence.
2. To provide an environment which promotes and reflects staff enthusiasm, enjoyment, involvement, and loyalty.
3. To stimulate, and provide support for, Communication Clinic staff to reach their own level of professional and personal excellence.
4. To protect confidentiality of client and staff information.
5. To conduct ourselves in a manner which exceeds customer expectations and ensures customer satisfaction.
6. To motivate people to help themselves and provide information to improve communication.
7. To provide services to our clients which are comprehensive, individualized, caring, and focused on ability rather than disability.
8. To encourage clients to reach their communication potential at their own stimulated pace with positive self-esteem.

9. To provide consultation services, screenings, and public information in order to identity and prevent communication disorders.
10. To deserve and present a positive public image.
11. To inform the public, clients, and associated businesses of Communication Clinic's scope of services, distinctive approach, and business practices.
12. To be sensitive to community needs, provide fair and equal treatment, and recognize and expose unethical practices.
13. To advance the field of communication arts and sciences through research, innovation, excellence, and participation in professional organizations.
14. To receive sufficient return on effort in order to finance attainment of these objectives and provide rewards for those pursuing these objectives.

CHAPTER 6

Establish the Structure

Defining the Organization [Subsection 2.2, page 89]

The reporting relationships (hierarchy of management), authorities, and responsibilities should be evaluated and (re)defined by the Quality Council. The organizational definition needs to identify positions and their general functional responsibilities and authorities. This may be documented through a combination of a hierarchical chart identifying the reporting relationships and a list of positions identifying the functional responsibilities and authorities. In addition to the management position names, the hierarchical chart may include the names of the managers; it may also include the names of the other people (the larger the organization, the more helpful this becomes).

The organizational definition should not contain detailed responsibilities or authorities. The organizational definition provides the basis for assigning detailed responsibilities and authorities as the details are (re)developed.

The balance between stability and flexibility in the organizational definition is very important. Frequent changes in the reporting relationships and functional responsibilities tend to cause major breakdowns in implementation of quality system due to confusion over responsibility and authority; and can cause a major drain in resources due to the need for constant change in the documents which define the quality system. Flexibility is essential in effectively responding to changing situations and maximizing the utilization of individual capabilities.

The following are important considerations on development of the organizational structure.

1.   The organization should be as flat as feasible. Extensive layering in the organizational structure tends to reduce the flexibility of the organization, increase the cost of operation, and insulate upper management from the people who actually produce the products and services for the organization's customers.

2.   Clear accountability should be established for all activities of the organization. Functional areas and administration and production processes should be identified and defined and responsibilities assigned to specific positions in such a way as to leave no question as to the owner. The importance of this consideration increases as the number of organizational units involved in a process increases. Clear functional accountability also helps minimize destructive turf battles as the organization grows.

Skills, Tools, and Information

Skills, tools, and information are essential elements in the success of an organization. (As used here, the term *tools* may pertain to anything which is used in accomplishing an objective. It may include skills and information, but never people.)

Mechanisms should be developed and/or enhanced to ensure that people have access to and use effective tools, and are skilled in the use of these tools. These mechanisms should also ensure that appropriate information is captured and used. The question is "How do we accomplish this?" Consider the following.

> Improvement requires change, but change does not guarantee improvement. Change leads to improvement when it results from the application of new knowledge. New knowledge is obtained through the process of

learning and dependent upon the effectiveness of the learning process. Therefore, the faster we learn, the faster we improve.

There are two distinct ways to learn:

1.   Learning from the acquired knowledge of other people, such as through reading a book or procedure, participating in a seminar or class, on-the-job instruction, or watching a training video. To be effective, mechanisms in this category must include a close link between on-the-job application and receipt of the information, and allow people time to reflect upon and share their new knowledge. One example of this results-oriented training is the teaching of team problem solving techniques as the team is working on their first project. This type of learning must come first; through it, new tools and skills are obtained and new ways of thinking are established.

2.   Learning through the creation of new knowledge/understanding, such as developing and testing new theories or the application of analytical/evaluative tools to information (data). To be effective, mechanisms in this category must ensure that the people continue to obtain new tools, are skilled in the use of the tools, and are allowed access to the needed information. The mechanisms must also ensure that the tools are used and the people in the organization are allocated time for these efforts (skills are obtained and enhanced by constant practice).

Effectiveness of the learning process is also affected to a large extent by the following:

- The understanding that there is a better way to do everything and our job is to discover and improve upon that better way (what is good enough today, will not be tomorrow).
- The organization's culture and environment must support efforts to learn and improve. This includes prompt constructive feedback and reinforcement.
- Learning must be approached in a planned manner to facilitate achieving an optimum rate of individual growth. For example, the learning of new skills must be built upon current knowledge and experience. It is also important to understand that different people learn in different ways, and to gear individual plans accordingly.
- New knowledge, or lessons learned, must be instilled in the fabric of the organization.

Communication and Information Systems

Mechanisms for effective communication should be established to ensure communication between organizational units and levels, and between the organization and its customers and suppliers.

Mechanisms for communication within the organization should ensure:

- Regular communication between and among management and their employees in the form of meetings, briefings, and documented information (or through information technology facilities),
- Management is informed of problem/concern occurrences and corrective actions, performance measures and results, financial status, etc.,
- Everyone in the organization is aware of the organization's customer commitments and goals, and of their part in fulfilling those commitments,
- Everyone in the organization is periodically informed of the organization's plans and the status of implementing the plans.

One of the most effective, broad ranging, and essential systems for communication deals with the collection, analysis/evaluation, summarization, and reporting of quality related information (this system is referred to as the Quality Monitoring System in this book). Another essential communication system, closely linked to the Quality Monitoring System, is one for monitoring/reporting the status of major activities (this system is referred to as the Project Monitoring System in this book). These performance monitoring systems are discussed later in this chapter.

Structured teams can also be an effective communication mechanism. For example, interfunctional and organizational interface teams provide an excellent mechanism for managing communication between functions and organizational units.

Mechanisms for communication with customers and suppliers should ensure that the organization is listening to them and keeping them informed. For example, the mechanisms should ensure that:

- Customers are aware of the organization's products and services, and understand the price structure and interface agreements,
- Customer's problems/concerns are promptly and effectively resolved and that the customers know where and how to express their concerns,

- Customer input is routinely obtained to evaluate and improve understanding of customer needs and satisfaction levels,
- Suppliers are aware of the organization's needs, involved in establishing schedules, and promptly informed of problems,
- People who interface with customers and suppliers are effective communicators.

Structured Teams [Subsection 5.7, page 102]

The term *structured teams* refers to teams which have been formally established, have a trained team leader, and are comprised of team members who are trained in team problem solving process and techniques. The word structured also infers that there is a program for overall management of the teams. The following are examples of structured teams.

- The Quality Council (discussed in Chapter 4)
- Quality Circles are probably the most prevalent example of a structured team program, and tend to be the first structured team program developed by manufacturing based organizations. Quality Circles are usually production level teams which are self-directed,

work on problems within their individual production areas, and maintain essentially the same team members over an extended number of projects.

- Quality Improvement Teams tend to be professional and managerial level teams which operate similarly to Quality Circles but with substantially more authority. These teams may be formed to address particular product/service lines, functional and interfunctional areas, organizational interfaces, etc.

- Interlinked management teams where each person is a member of the team led by their supervisor and the leader of the team in which their employees participate. (The success of this form of team has a greater dependence on the leadership ability of the managers/ supervisors. It can result in an organization of effective team oriented work units and provide an effective management communication process. It can also result in the destruction of all team programs.)

- Another variant of structured teams are teams which are formed at management direction to address a specific problem or project. Although the use of problem/project teams is a common practice, the other structured team approaches are an important ingredient in Total Quality management.

The following is an outline of a process for developing structured teams.

1. Select one of the trained and experienced team leaders who have peer respect to be a team facilitator and provide this person with facilitator training (or select the Quality Council facilitator). This

person should be assigned responsibility for developing and managing the team programs; and he/she should report to the Quality Council either directly or through one of the council members to ensure that their primary focus is on developing team programs.

2. Select one or (preferably) two forms of structured teams and develop a program for each form selected. Each program should define the process for selecting the team members and leaders, training requirements, extent of autonomy, team progress tracking mechanisms, and reward/recognition considerations.

3. Initiate pilot teams. In general, it is advisable to initiate three teams for each of the structured team programs (maximizes likelihood of having at least one very successful team in each program). All teams should be formally facilitated, especially in the beginning. As a minimum, the facilitator should periodically observe team activities and provide counsel to the team leader; the facilitator should also encourage and initiate feedback from the team members.

4. Revise the programs based on the result of the pilot teams and train additional team leaders and facilitators. (Team leaders and facilitators should form a cohesive group and share information on successful efforts as well as methods which appear to fail.)

5. Expand the number of teams as team leaders and facilitators become available (include a team of leaders and facilitators which will focus on improving the team programs).

6. Expand the number of programs as the orginzation becomes increasingly involved in team activities.

Note: Remember, team participation should be voluntary.

SO Note: Shoot for at least one program with two teams using different leaders who may be members of the Quality Council. Let the leaders function as facilitators for each other's team.

Individual structured teams are generally formed with one or more of the following basic objectives:

- Improve production, administration, and management processes (this is the predominant basic objective, especially during the first few years of structural team operation, because of the tremendous and prompt benefits gained by improving upon current processes)
- Improve product/service design
- Create new products, services, or processes
- Manage major projects.

No matter the reason that a team is formed, the team members must understand that they represent all owners of the process or project on which they will working, and are themselves owners of the process/ project. Of course, management must first accept and always demonstrate this principle. Management must also provide prompt constructive feedback to the teams, recognition for team participation, and rewards/recognition for successful team efforts.

Performance Monitoring Systems

Monitoring systems are used to collect and manage information related to quality and the pursuit of goals. They are essential tools in quality management which:

- Promptly provide each person and organizational unit with feedback on the quality of their products and services such that they can take corrective action (adjust their activities) on a real-time basis (in other words, *ensure* that the quality of their productive efforts is what it should be)
- Provide information for assessing the status of quality and progress toward goals within the organization (in other words, *assure* that the quality of productive efforts is what it should be).

In this section, we will consider two monitoring systems: The Quality Monitoring System and the Project Monitoring System. In each of these, the information is entered into the system for or by the person who uses it to monitor their own work. This information is summarized and supplemented until it reaches the top manager in the form of key status indicators of quality within the organization and of progress toward major goals.

In the Quality Monitoring System, detailed information on quality characteristics is collected by and for the people who use the information in monitoring, controlling, and improving their own products and services. The characteristics monitored and the measuring/monitoring instruments used to collect and evaluate this information are generally under the control of the initial users of the information, and are selected/ designed to best meet the initial user needs. When summarized, combined with related information from other areas of the organization, and converted to suitable form, the information enables busy managers to become adequately informed as to quality performance and trends without becoming heavily involved in production operations. It also enables management to provide feedback to their organizations and to periodically inform everyone of the organization's status with regard to quality. At the information's most refined level, it allows comparison of performance against the organization's strategic goals and establishment of new goals.

Note: Do not be mislead by the general words in the above paragraph such as evaluate, summarized, combined, and converted. The use of statistical methods during these activities is essential to the success of the system. Reports to management must (in practically every case) be based on, and reflect the results of, statistical evaluation of the data. Only through the use of statistics can management effectively minimize both types of error in decision making: (1) taking corrective action or providing rewards/recognition when in fact nothing has changed, or (2) not doing it when changes have actually occurred.

Examples of the raw information include product/service inspection results, details on returned products and customer complaints, repair data, process yield, participation rates, training test scores and carryover rates, employee turnover, financial data, quality costs, etc. Examples of the information at it's most refined level include history plots of percent on-time delivery, percent error-free products/services, product reliability, customer satisfaction, Total Quality costs, return on investment, etc.

The Project Monitoring System is used by management to monitor and direct important projects and activities within their area of responsibility. Each person in the organization periodically (generally monthly) provides a brief report to their immediate manager on the status of their major activities and progress toward their goals. People with lead responsibility for group projects also include a brief statement on project status. Each manager compiles and summarizes the reports from their employees, adds their status statements, and provides the report to their manager. At the top of the organization, the report addresses the status of critical projects and significant activities.

The overall Quality Monitoring System and the Project Monitoring System are addressed in Section 3 of the "Quality System Desk Reference" (page 91). Collection of specific data is addressed throughout Sections 4–13 of the desk reference.

The Quality Council should ensure that performance monitoring systems are established to collect quality related information and project/activity status information. In general, the Project Monitoring System is fairly easy to institute and will be a relatively stable system. The Quality Monitoring System will be very dynamic because the set of quality characteristics monitored will change as products, services, and processes are created and changed, and new characteristics will rise to the top of the problem list as old problem characteristics are brought under control.

Strategic Planning [Subsection 2.5, page 90]

Planning is essential to effective management of resources, quality, and growth. Once each year the Quality Council should develop a plan for operation, improvement, and growth. It should address the following three to five years with the majority of the emphasis on the next twelve to eighteen months.

In developing the plan, the council should consider the needs, aspirations, market forecasts, and resources from all aspects of the organization, including the following:

- Equipment replacement and upgrade,
- Development and enhancement of the quality system,
- Market consolidation/expansion,
- Product/service development and major modifications,
- Major training programs,
- Major problem recurrence control activities,
- Major changes in resource allocations and needs,
- Major quality improvement efforts—breakthrough management [Subsection 4.7, page 98].

The planning process should result in a set of one to five year objectives with major milestones, resource allocations, and responsibility assignments, including a budget for the next fiscal year.

The planning process should involve as many people in the organization as possible and may be as follows:

1. Request input on needs, aspirations, and market forecasts from the organization. This information should be generated from all segments of the organization (use of team techniques for idea divergence/convergence will substantially improve the quality of the information) and compiled as it moves up the organization. It should include the identification of priorities and justifications.

Note: Primary information sources used during this process are the Quality Monitoring System and Project Monitoring System discussed above.

2. Compile all information into a single draft plan.
3. Project financial, material, and human resources to ensure achievability of the plan. Assign responsibilities for leading each project.
4. Distribute the plan throughout the organization for review and comment.
5. Revise the plan and redistribute it to obtain organizational unit and personnel assignments for implementing the plan (to the extent feasible, assignments should be made on a voluntary basis).
6. Revise the plan to include primary responsibilities, and issue the plan and budget for implementation (generally, many of the projects have already been initiated by the time the plan is issued).

SO Note: Everyone in the organization should participate in the generation of ideas, prioritizing, review and comment, and responsibility assignment activities.

Progress toward achieving the milestones should be monitored throughout the year and adjustments made as needed in priorities, resource allocations, and milestones. During the first several years, the planning and monitoring processes will evolve dramatically, as will the types of goals selected and the reporting mechanisms.

Refer to Appendix II for examples of annual strategic plans (page 0000).

Establishing the Procedure System [Subsections 1.3 and 4.2, pages 88 and 96, respectively]

Procedures provide the primary mechanism for management to ensure that there is a common understanding of the organization's focus and quality system, and that the organization's activities are in control. The procedure system provides the controls over the procedures.

Note: For the sake of brevity, the term *procedure* as used here pertains to all documents which prescribe guidance or control activities, including policies, programs, procedures, guides, and detailed work instructions.

The Quality Council should place considerable attention on developing an effective procedure system. A decision to make major changes in the system once a substantial number of procedures have been issued should be given due consideration. Major changes could result in considerable attitude damage in those who have to rewrite otherwise good procedures, and require considerable resources.

The following are important considerations on development of a procedure system.

1.   While procedures provide the primary mechanism for ensuring that activities are in control, they can also be the primary force inhibiting employee empowerment. Procedures best serve the organization when written to *help* (empower) the people using the procedure to accomplish their activities in the best possible manner. In general, this means that procedures should lean toward providing a guide to the specific actions; and restrictive details should be included only to the extent that success of the activity depends on a specific sequence of precise actions.

2.   Organizations which have little or no experience with procedures should be cautious in introducing procedures to their employees as people from many cultures initially perceive procedures as a reduction in their personal freedom and an attack on their workmanship. One approach to minimize this reaction is to develop an interim manual of policies and program descriptions which provide brief guidance for important activities. In this case, the emphasis should be on those activities which have a primary impact on employee and customer satisfaction. This manual should be developed using the procedure system development process outlined in this chapter and Chapter 7. The Quality Council should reinitiate the development process to replace the interim manual only after the employees are consistently using the manual, the additional control provided by more comprehensive procedures is deemed appropriate by the Quality Council, and the employees have accepted the idea of procedures. Appendix III (page 147) provides an example of an interim manual. A complementary approach to minimizing adverse employee reactions is addressed in paragraph 3 below.

SO Note: Unless special circumstances exist, this "interim" manual may be all that you need for the majority of your activities. For those unique activities which deserve greater control, the applicable policy/program description may be expanded or another manual may be initiated to supplement the policy and provide the additional detail (such as to address accounting procedures). Although this approach is generally better for plugging a few holes, it can cause severe problems if used extensively. (It is like patching potholes. The patches may reduce specific problems, but the greater the number and size, the bumpier the road.)

3. As mentioned above, some people resist the use of procedures. Whether or not resistance is an established fact in an organization, an excellent approach to overcoming this resistance is to establish the realization that the procedures belong to the people who use them and exist to help the employee ensure customer satisfaction. This can be accomplished by consistent employee involvement, and eventually leadership, in the evaluation and redevelopment of their activities and the corresponding procedures. Let me reemphasize that (re)development of the procedure system must not proceed until the employees recognize that the culture is changing for the better. Remember, the foundation (Chapter 5) comes before the structure (Chapter 6).

4. To the extent feasible, the number of procedures should be held to a minimum; procedures should be of reasonable length, and be self-contained (all requirements and guidance needed to perform a task should be in one procedure, not dispersed throughout the procedure system).

5. The procedure system should be comprised of one manual or a hierarchy of manuals with one manual at the pinnacle (top) of the hierarchy.

6. In a hierarchical system,
- A common format and numbering system should be used throughout the hierarchical system,
- The pinnacle manual should contain the statements of mission, distinctive competency, and objectives, and prescribe the guiding policies and programs for the organization,
- Each manual in the second level should address a defined segment of the organization or functional area, provide for implementation of the top manual within that defined segment/area, and prescribe additional detail for the manner in which that segment/area will operate,
- Each manual in subsequently lower layers of the hierarchy should address smaller and smaller segments/areas and provide increasingly more detail.
- To minimize the number of procedures, procedures which are to be implemented throughout the organization or unit should be implemented directly from the uppermost manual in the organization/unit (e.g., one procedure in the pinnacle manual for handling of problems/concerns which is used by everyone).

There is another consideration which may affect your organization, especially if you operate in a government regulated industry or provide products/services to the federal government. That is, a requirement for the existence of a manual documenting the organization's quality assurance programs (frequently referred to as a QA Manual, QC Manual, or Quality Manual). In general, use of a Quality Manual is most effective when it is treated as a source of requirements for the overall management system, or as a description of how the management system ensures compliance to a set of specific requirements or for a specific project. In any case, the preferred alternative should be to seek approval of the customer or regulatory agency to identify your pinnacle manual as the Quality Manual.

In developing the procedure system, the council needs to focus on developing a draft index of the pinnacle manual and identifying second level manuals, as appropriate.

The following is a process for developing the procedure system which may be used.

1. Outline the general structure of the procedure system. For example, will it be a single manual or tiered system of manuals?
    *   A single manual should be used when the organizational structure and processes are simple, and multiple organizational units are involved in most administrative and production processes. (Although the company used for the examples in this book meets these criteria, a tiered system was chosen. This decision was based on an industry requirement that one specific segment of the organization be clearly directed by an industry certified individual.)
    *   When a hierarchy of manuals is to be used, identify the scope of each manual in the top and second tier of the hierarchy. Each manual in the second tier may address strongly interrelated functional areas (as an example of this, see the procedure system outline example which follows at the end of this section) or based on the organizational structure.
2. Develop a draft index of the pinnacle manual with the scope of each procedure outlined for use in guiding manual development. This index needs to be logically organized to facilitate location of information in the completed manual. The council should clearly understand that this is a draft index which will be revised as the procedures are developed. As a minimum, the index should provide for inclusion of:

- The statements of ultimate focus developed in Chapter 5,
- The organizational definition,
- Principal policies and requirements which will direct all aspects of the organization including control of the procedure system, and
- Procedures which are to be directly implemented by the entire organization. (In some tiered procedure systems where the second tier is organized by function and applied to all organizational units, these procedures are contained in second tier manuals.)
3. Draft a procedure for definition and control of the procedure system. This procedure will be used to guide development of the initial procedures and should be among first procedures to be issued. As a minimum, this procedure should:
- Define the scope of the pinnacle and second tier manuals, and may identify all manuals in the procedure system,
- Prescribe the general requirements for format, contents, development, review, approval, distribution, and revision of procedures; and prescribe the details of the process for at least the pinnacle and second tier manuals (this eliminates the need for each manual to have a procedure on procedure control).
- Require that each new and revised procedure be promptly read by those who must implement them, and may provide a mechanism for ensuring that this responsibility is fulfilled.
The procedure may:
- Provide guidance on procedure drafting to ensure consistency and adequacy, and
- Provide guidance on control of terms which will be used by the whole organization.

Below is an example of a procedure system outline and draft pinnacle manual index. These examples were developed using process steps 1 and 2 above. For brevity, much of the detailed procedure scope information has been deleted from the index.

Refer to Appendix IV (page 155) for a comprehensive example of a procedure on procedure system and control, and a guide to drafting procedures. Control of the "interim" manual of policies and program descriptions is included in Appendix III.

Policy and Programs Procedure Manual

Treatment and Documenta- tion Procedures Manual	Office Procedures Manual
Statements of Mission, Distinctive Competency, and Objectives	Statements of Mission, Distinctive Competency, and Objectives
Treatment Procedures	Office and Accounting Procedures
Interface Procedures (copies from other manuals)	Interface Procedures (copies from other manuals)

Example
Policy and Programs Procedure Manual Index

Procedure Number	Procedure Title and Scope
0.1	Statements of Mission and Distinctive Competency
0.2	Objectives
0.3	Programs Overview

STAFF GUIDELINES

1.1	Quality Focus
1.2	Image, Attitude, Ethics
1.3	Management
	a. attitude: lead by example, deep trust in people
	b. associate guidance
	c. management style
	d. meetings and status reporting
	e. organization and responsibilities
1.4	Associate Responsibilities

SERVICE PROGRAMS

2.1	Programs Management
2.2	Treatment
	a. policies
	b. therapy procedure guidelines
	c. documentation procedure guidelines

2.3   Client Satisfaction Management

2.4   Client Administration

SUPPORT PROGRAMS

3.1   Social Responsibility

3.2   Information Dissemination

3.3   Procedure System and Control

3.4   Documentation Control

a. filing system

b. forms

c. records

d. computer files

3.5   Financial, Contractual, and Legal Management

3.6   Procurement Control

3.7   Equipment and Materials Control

3.8   Quality Concerns and Corrective Action

3.9   Evaluation, Innovation, and Improvement

a. financial and quality audits

b. program evaluations

c. research and team programs

d. quality measures

3.10   Human Resource Management

3.11   Staff Benefits

a. salaries and mileage reimbursement

b. promotions

c. raises and bonuses

d. educational and conference reimbursement

3.12   Growth Management

a. annual plan

b. beating the bushes

c. development of new facilities

d. development and change control of service lines

Critical Policies and Programs

Policies and programs which are critical to operation or provide essential quality monitoring and improvement mechanisms should be considered part of the structure. Such policies and programs should be promptly (re)developed and documented. The (re)development and documentation of the remaining policies and programs are addressed in Chapter 7.

It is important to understand that the process of procedure development is more important than merely documenting policies and programs. It should be a process for evaluating current programs, policies, requirements, and processes to identify weaknesses, improve efficiency and effectivity, and adjust the balance between requirement and guidance.

The Quality Council should identify critical policy and program areas and ensure that these areas are the first to be addressed in procedures. Two areas which should be at the top of the list are:

- Procedure control program [Subsection 4.2, page 96]
- Problem/concern program [Section 11, page 118] (Provides the structure for control of problems/concerns). *Note:* Refer to Appendix V (page 171) for an example of a procedure on problem/concern control, corrective action, and the associated processes. You may note that this example also provides for the identification, evaluation, and enhancement of good practices.

Other areas which may be included in the list are:

- Document control and records management [Subsection 4.3, page 96; Section 10 page 117]
- Customer satisfaction management program [Section 13, page 124]
- Career planning and training policies [Section 5, page 100]

In the interest of establishing guidance for use during the development of the implementing procedures, the Quality Council may issue a temporary procedure or memorandum to define policies and outline programs. This is also a good method for testing the need for and adequacy of new policies and programs before expending resources in development and implementation of new procedures in the policy area. Examples of areas for which this approach may be best suited include the structured team program, communication and information systems, and strategic planning.

CHAPTER 7

Completing the Basic Quality System

General Overview

At this stage of the transformation, excitement will be clearly building throughout the organization. More and more people are becoming involved. The organization will begin to witness the benefits of the transformation, including more satisfied workers, substantial reduction in the cost of operation, and improvement in the quality of the products and services provided to the organization's customers. This is a time of fast-paced change. It is a time when the opportunities are boundless and the risks of failure are equally as great. You might say this stage contains the hump in the transformation process.

The Quality Council must simultaneously manage change in diverse aspects of the organization. Extensive planning and monitoring of the transformation is now critical. This may at times seem like an unmanageable burden. However, if the structure was properly laid, the Quality Council will be in a position to manage the changes effectively. The following are aspects of the structure which may be critical during this stage of the transformation:

- The annual strategic plan identifying major goals (changes), milestones, and responsibilities, with lower level implementing plans for each major element of the strategic plan,
- Well-defined structured team programs with responsibility assignments for managing the programs,
- Well-defined training needs and plans with responsibility assignments for implementing the plans,
- Established communication systems with responsibility assignments for managing the systems.

Broad Focus

During this state of the transformation, the Quality Council must maintain a broad focus. No aspect of the organization or the transformation process may be overlooked. However, there are specific aspects of the transformation process which are deserving of special attention.

- The skills of everyone in the organization and the tools used within the organization should be enhanced.
- The use of structured teams should be expended drastically (sixty percent participation is a reasonably achievable goal). This effort will probably be the greatest source of improvement for efficiency, productivity, product/service quality, and skill and tool enhancement.
- The strategic planning system should be expanded and refined. As mentioned above, extensive planning and monitoring is critical during this stage of the transformation. This system will be the primary tool for directing the changes. As such, it should be expanded into a tiered structure of plans. Each manager should have a strategic plan for his/her own area of responsibility, and each of these strategic plans should contain elements which support the plans at higher management levels within the organization.
- The Quality Monitoring System and Project Monitoring System should be expended and refined. To the extent feasible, these systems may be substantially automated. Use of automated data collection systems will increase the amount of information available and reduce the effort required to generate the summarized reports. For

example, with all Project Monitoring Reports on the same word-processing system, the paragraphs addressing the status of annual strategic plan projects can be readily compiled into one report by the Total Quality Coordinator.

- All current policies, programs, requirements, and procedures should be evaluated to identify weaknesses, improve efficiency and effectivity, and incorporate the organization's quality principles and statements of focus into the fabric of the organization.

As all but the last aspect was addressed in Chapter 6, the remainder of this chapter will be devoted to (re)developing procedures throughout the organization.

Completing the Procedure System

This effort is accomplished by systematically (re)developing procedures to address all aspects of the organization and all applicable quality system elements identified in the "Quality System Desk Reference." Procedures which already exist should be reevaluated to ensure that they are in agreement with the new focus of the organization, are effective, and logically fit into the overall procedure system.

Note: Those developing the "interim" manual of policies and program descriptions will use this same process, but will be developing short guides which are compiled into a manual instead of complete procedures. See Appendix III (page 0000) for an example.

On first reading, many people feel that they do not have sufficient knowledge to undertake this endeavor. Cancel this thought; it will only interfere with the pursuit of excellence. Remember, by the time you actually reach this point in the transformation process, you will have obtained considerable knowledge and experience in quality management. In fact, because of your personal drive for excellence, you may have a far greater understanding than many "professional" quality managers who were driven by their "bosses" or are in the profession because of job title. Beyond knowledge and experience, there is another management attribute which is essential: common sense; and if you are seriously interested in the pursuit of improvement and results, you probably have an ample supply of common sense.

This effort may take several years; in fact, a quality system which is developed slowly and with considerable thought tends to be significantly better than one which is pumped out. So do not get discouraged—get organized.

The Quality Council should coordinate an overall plan for development of needed procedures based on a top-down approach, and monitor progress on implementing the plan. This plan should result in the expansion of the procedure system outline and draft pinnacle manual index. The plan should address the use of structured teams in (re)developing the programs and procedures, and should be addressed in the annual strategic plan.

Although all quality system elements should be addressed during the planning, the following elements need to be given special attention as they will be used for ongoing control and improvement:

- Communication systems
- Quality and Project Monitoring Systems [Subsections 3.5 and 3.6, pages 93 and 94]
- Strategic planning [Subsection 2.5, page 90]
- Structured teams [Subsection 5.7, page 102]
- Breakthrough management [Subsection 4.7, page 98]
- Human Resource—functional definition and indoctrination/training [Subsections 5.2 and 5.3, page 100]
- Quality audits and management reviews [Subsections 3.2 and 3.3, pages 91 and 92]
- Procurement program [Section 8, page 109]
- Process monitoring, control, and improvement [Subsections 9.3 and 9.5, pages 114 and 116]
- Cost/revenue management [Section 12, page 121]
- Customer feedback and customer concerns [Subsections 13.3 and 13.4, page 125]

The plan should be flexible to allow for procedure system improvements. Inevitably, as procedures are developed, detailed evaluation of the programs and processes will result in changes to the procedure scope; and these changes will impact the other procedures. Therefore, periodically as procedures are developed, the plan should be revised to reflect the changes (including, elimination of some planned or issued procedures, and revision in the scope of others).

Plan for realistic and phased schedules such that the most important programs and processes are addressed first and the least important last, and the individual programs developed from the top down (general procedures first and detailed implementing procedures last). This has a benefit of providing time for the organization to become familiar with the use of procedures before resources are wasted on very low priority procedures which are frequently unnecessary. It also has the benefit of ensuring that detailed procedures are not generated for new programs until experience is gained.

CHAPTER 8

Living with Total Quality

Fait Accompli

The time will arrive that your Total Quality culture and systems appear to be firmly established and self-perpetuating. Time to breath easy. Or is it? Fait accompli (an accomplished fact, presumably irreversible) was chosen as the theme for this section because it expresses the customary sentiment of the Quality Council and others in the organization at this time. Unfortunately, it can be an inappropriate sentiment and one which leads to a setback in the path to achieving excellence. Fait accompli implies an irreversible fact. This is far from the truth in this instance.

In a sense, fait accompli applies only to the extent that the members of the Quality Council continue to constantly:

- Live by and exemplify the quality principles, and pursue the organization's statements of ultimate focus
- Maintain focus on improving quality of work life, skills, and tools

- Maintain awareness of the customer's needs and satisfaction, and ensure that the needs are fulfilled and satisfaction is enhanced
- Maintain awareness of the attitudes expressed within the organization, pursue attitudinal improvement, and take corrective action whenever and wherever attitudes take a nose dive
- Pursue greater levels of, and more meaningful participation
- Ensure ongoing control and improvement in all aspects of the organization.

In short, fait accompli applies when management continues on its course of leading the pursuit of excellence.

Now lets discuss strategic quality management and the two basic aspects of an effective management system (control and improvement), and see how they relate to living with Total Quality.

Strategic Quality Management

Strategic quality management occurs when the supporting quality systems have been established and are used to plan, control, improve, and monitor quality. It is a process for steering into the future and maximizing competitive quality advantage. It links all people and activities to the objectives which they support, and ensures that action plans are established at all levels and in all areas of the organization.

Strategic quality management uses the Quality and Project Monitoring Systems, quality cost management, problem/concern management, and audit and management review programs to monitor performance against stated objectives, policies, procedures, and goals. It uses the strategic planning, breakthrough management, structured teams, problem/concern recurrence control, process management, procurement, and human resource management programs to direct and reach new higher quality performance levels. It uses the marketing and product/service development and change control programs to ensure quality is designed into the product/service. It also uses the customer satisfaction program to ensure that customer expectations are exceeded.

Making the pieces fit together effectively will take time, patience, knowledge, and experience. But once you have accomplished this, your organization will have succeeded in achieving a Total Quality culture.

Culture—Pursuit of Excellence

Ongoing Control

As the quality system is developed and documented, control consists of ensuring that the quality system is implemented. This is achieved by:

- Management consistency in demonstration of the quality precepts,
- Devoted attention to attitude development, motivation, team programs, effective communication, and quality training,
- Devoted attention to ensuring that everyone in the organization understands the organization's focus, and quality system requirements and guidance,
- Consistent management attention and insistence that everyone understand and comply with the procedural requirements and guidance,
- Maintenance of the quality system such as prompt revision of procedures when procedural deficiencies are identified or the organization is altered,
- Ongoing strategic planning,
- Performance monitoring,
- Management reviews, and quality system and financial audits,

- Incorporation of lessons learned into the fabric of the organization (such as by revising the procedures and sharing new skills).

Ongoing Improvement

People become more and more excited as the roadblocks are removed and they realize that they can have an impact (people want to achieve excellence, mediocrity results from the system). One idea stems from the last and improvements build upon each other. People gain greater understanding of the processes in which they participate. Processes become more and more effective. Products and services become better and better, and lead into new products and services which become better and better.

Ongoing improvement is a relatively natural self-accelerating phenomena when management has provided the initial momentum, systematically constructed the tools and removed the obstacles, and continues to lead the pursuit of excellence. Ongoing improvement is the quality system aspect which perpetuates the sense within the organization, it's customers and suppliers, and the general public that the organization is fresh and vibrant.

Because of the intensity of ongoing improvement, many people tend to build an imaginary wall against further improvement. They theorize that there is a point at which improvement becomes unnecessary and impractical. Fortunately, this theory is invalid because of change—in the people involved, products and services, market place, technology, and customer's definition of quality. Unfortunately, an imaginary wall is perceived as real by the people who created it.

The Quality Council should be constantly on the lookout for the construction of this wall; and wherever construction begins, the council should devise and institute a plan based on quality precepts for overcoming the misperception. The building of this imaginary wall may be encountered many times during and following the transformation process; but it will probably reach its pinnacle just as the Total Quality culture becomes a firm reality within the organization.

Ongoing improvement is achieved by continued and consistent:

- Focus on improvement in the production, administration, and management processes, and in the products and services produced for outside and inside customers
- Emphasis on and support for structured teams (the primary tool for achieving ongoing process improvement)

- Focus on identifying and meeting the customer's changing needs
- Focus on identifying strengths and weakness in competitors and other companies, and using this knowledge in the pursuit of excellence
- Emphasis on enhancing the tools and skills used by the organization
- Focus on improving the quality of life for all people within the organization, the customers, the suppliers, and society as a whole.

Where Do We Go From Here?

Achieving excellence in business is the title of this book. At what point does an organization achieve excellence? Some people may say that excellence is never achieved on our plane of existence, that it would be like reaching infinity. From another viewpoint, excellence is a way of living, not a destination; and it is achieved when we are continually and consistently pursuing improvement in every facet of our lives. Based on this second viewpoint, an organization has achieved excellence when it has completed the transformation, when Total Quality is indeed fait accompli.

Such an organization will have experienced significant progress since embarking on the Total Quality quest and accomplished previously inconceivable goals. The people within the organization will have experienced significant professional and personal growth, and will have become a healthy and vibrant family. In addition, quality of life among the customers and suppliers will have improved, and the environment and society as a whole will have gained.

The organization will be focused on a common mission. The management will be experienced leaders. Everyone will understand the magnitude of power available when each person has the opportunity to make meaningful contributions to the common mission. Continuous improvement will be a way of life; and one which will lead the organization to an ongoing series of new quests, and higher standards and expectations.

At this point in the organization's evolution, what will count to the customers and the members of the organization will not be achievements of the past, but the accomplishments of the future. By continued dedicated commitment to never-ending quality improvement, thirst for knowledge and innovation, and the power of teamwork, all involved can be assured of a future filled with new opportunities, challenges, ideas, and individual and organizational growth.

CHAPTER 9

Communities of Excellence

By implication, quality excellence includes the continuous pursuit of improvement, as an individual, organization, and community. As a person achieves success in their pursuit of quality excellence in the internal (human) domain and in their organization, the natural progression of improvement includes helping others to improve.

Communities of Excellence is a process whereby individuals and companies share their quality improvement expertise with members of the community and facilitate quality improvement efforts within the community. For example, in the Tri-Cities Tennessee Communities of Excellence project, teams from businesses, industries, government, services, and educational institutions generated over 14 million dollars of documented savings within the first few years.

The following is an excerpt from a presentation made at the ASQC Quality Congress. The presentation was entitled "Creating Communities of Excellence: An Overview" and was made by Wilbur R. Hutsell of

Tennessee Eastman Company (one of the principal companies involved in the Tri-Cities Tennessee Communities of Excellence effort).

Creating Communities of Excellence: An Overview

"The growth of the world economy and the declining U.S. competitiveness threaten the American standard of living and position of leadership in the world. Unless the U.S. can compete more effectively, the trends that are underway will continue—trade deficits, loss of jobs, and loss of whole industries. What is not clear is that the industrial decline affects more than those who work in obsolete 'smoke stack' industries; in fact, we are losing the high-value added emerging technologies that are our future. What is even less clear is that the decline is more pervasive than industry alone. It affects the whole fabric of the community. The issue is not just quality of goods and services, but quality of life as a whole.

" . . . The workers who live in a community that gives them poor services, poor education, and poor government are conditioned to work in a badly managed industrial system that produces poor quality products as well. They come to expect the phone not to work, the automobile repair to require several trips to fix and to cost several times what it should, and the sales clerk to be rude or to ignore them completely . . . The fact is that the community must be considered as a whole—it cannot be compartmentalized.

"There is a way out. *Quality* is the key to regaining the competitive edge, but it requires holistic quality, not just quality as related to the industrial scene . . . What is needed is a nationwide revolution of sorts— a quality attitude matched with quality action in all sectors of each community and in all sections of the country.

" . . . Each community can become an island of quality, a miniature Japan if you will, becoming a world leader in whatever it does. As these islands spread, one by one, collectively the challenge can be met.

"This then, is a vision of Communities of Excellence as a path to quality—a positive response to the challenge. It is a story of people in all walks of life working together in a very American way. In many ways it is a return to some very basic American values . . . We have seen some glimpses of it in the Tri-Cities region of upper East Tennessee, but we have by no means arrived . . .

In order to understand this vision, we must first understand that a community is an interacting group of people, with no set boundaries. Any way boundaries are defined, they are tentative and subject to change. A true Community of Excellence is not concerned with boundaries, but becomes an organic entity that is simultaneously unfolding possibility and reality built on the genius of the people of the community . . . However it is defined, a Community of Excellence must have the participation of all of its members, all dedicated toward total excellence and continual improvement.

"Total participation cannot be overemphasized. Lack of participation by one sector or strata leads to lack of interest, jealousy, and fear. The Communities of Excellence concept is based on widespread participation . . . But even when participation is truly happening, it is not enough to just have people getting together with no real purpose. Participation must be focused on *quality*—from the customer's perspective.

"Getting the community organized for participation and focusing the people on quality requires *leadership* . . .

1. Leadership *activates* the process by organizing, training, and providing the opportunity . . . Some up-front faith is required to invest some time, effort, and energy to start the process moving. This is very much like the farmer who buys seeds and plants them with

the faith that good things will happen later in the season. Once set in motion, the widescale participation focused on quality will generate an unbelievable amount of energy from the participants that is now just lying dormant. It will get people turned on and excited about what they do, from school teachers to janitors to telephone repair people . . .

2. Leadership *sustains* the process by keeping a *vision* in front of the people, recognizing their accomplishments, and acting on their recommendations . . .

"There are several aspects to the concept of Communities of Excellence—people are a very important one, but not the *only* one. It is essential to understand that leadership focuses on improving the system, not getting people excited. Dr. Deming is right when he decries management by exhortation. The people will, in fact, put forth extra effort as they become enthused and excited, but trying harder by itself doesn't lead to excellence—it leads to exhaustion and frustration. Real improvement will come about by working on the *system with the help* of the people, to bring about changes in the system . . .

"Consider the case of vehicle fleet maintenance in the city of Kingsport. This seems a rather mundane activity until you understand how their function fits within a larger system. Maintenance delays can affect the whole range of other city services—street paving, garbage removal, and even police and fire protection. By reducing downtime, improving communications, and improving the *system*, the people in fleet maintenance have increased quality and productivity in a host of other functions. In addition, the realization of how they contribute to the larger system has increased the pride of the mechanics and broken down barriers between them and other city functions . . .

"Communities of Excellence . . . are using systems-type thinking to organize their entire community effort, building on the bedrock principles put forth by Deming, Juran, and others. The real Communities of Excellence exist when all the pieces of the community focus on quality, using the quality tools and systems-type thinking. However, this is vision toward which we are moving, not the starting position. Each community will have to start with what it has, build on it, and nurture it until critical mass is achieved . . . "

(Note: For brevity, discussion of the actual steps taken in Tri-Cities Tennessee to set the process in motion has not been included.)

"What have we learned from our experiences? Four things stand out:

1. Quality principles apply to the *whole* community.
2. The power of teams . . .
3. Everybody can use his mind as well as his hands to bring about improvement. This is an *affirmation* of the American *creative genius* and is a *competitive advantage* to us to be the best!
4. The energy to improve is there already. This is an affirmation of faith in the American worker . . .

"Although one can list characteristics or cite examples, the Community of Excellence is much more. It must be *experienced* to really be understood."

Through sharing of our pursuit for excellence, we help others to pursue human excellence. We contribute to the future of our community, nation and world. We enhance value and fulfillment in our lives and the lives of all inhabitants of our world. In short, we find and fulfill our purpose.

For more information on Communities of Excellence and the process for initiating the program, contact the Chamber of Commerce in Kingsport, Tennessee, your local chapter of the American Society for Quality Control, or Jackson Community College in Jackson, MI.

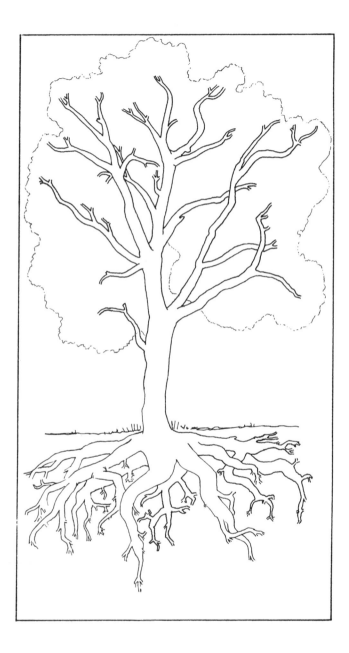

PART 3

Quality System Desk
Reference

This reference manual consists of this introduction and 13 sections, with each section addressing a different functional area of the quality management system.

The organization of this reference manual is a logical grouping of the information which was chosen to facilitate an understanding of major functional groupings and assist the reader in realizing the holistic nature of an effective quality system. Examples have been minimized for the sake of brevity. (The process for development and improvement of the quality system and examples were discussed previously in this book.)

If this is your first review of the quality system desk reference, your general understanding may be facilitated by periodically referring back to the following Table of Contents and making brief notes on the individual subsections and their relationship to the whole system.

During the initial review of this reference manual, it is important to remember that an effective quality system is built and evolved over time, in step with the evolving attitudes of the organization's people, customers, and suppliers. Quality system development starts by focusing on selected aspects of a few functional areas. Quality system improvment consists of addressing new areas and enhancing aspects of currently implemented elements.

In other words, do not allow yourself to become overwhelmed. All you need is a general understanding of quality precepts and the quality system scope and relationships. The details will be addressed a little at a time as your management system evolves; and, at those times, the details may be obtained from the applicable sections of this reference manual.

Special Word Usages

As stated in Chapter 1, the use of industry specific terminology has been minimized. Where questions arise related to the meaning of specific words, a standard collegiate dictionary should provide the needed guidance.

The following additional word usage clarifications are provided to assist in interpreting and implementing this reference manual.

Should and May

The words *should* and *may* have been selected to convey the advisory nature of this guide. However, in any specific application, a document based upon these guidelines should clearly indicate those provisions of the management system that are obligatory through the selective use of the words *shall* or *must*.

- The word *should* is intended to stress the importance of the provision and indicate that very careful consideration is warranted.
- The word *may* is intended to identify those provisions which would provide additional benefits to the user.

Periodic

The word *periodic* and other references to frequency are used to indicate that the associated activity should be performed at some stated frequency. During implementation of this reference manual, the organization should replace each stated or implied use of the word *periodic* with a specific frequency as appropriate to the circumstances. For example, "Plans should be periodically reviewed and updated to reflect changing situations" may translate into "Plans will be reviewed and updated quarterly" or "Plans will be reviewed and updated every month" depending on the importance of the plans and the rate at which the situation tends to change.

Quality Characteristics and Requirements: Qualitative vs Quantitative

Products, services, and their production and delivery processes have qualitative and quantitative characteristics.

* Quantitative characteristics are measurable. For example, waiting time, delivery time, completeness of service or order, accuracy of billing, diameter, temperature.
* Qualitative characteristics may require subjective evaluation. For example, courtesy, responsiveness, comfort, aesthetics, dependability.

Requirements provide the basis for judging the acceptability of the characteristics. Service requirements relate to those characteristics directly observable by the customer. Product requirements relate to those product characteristics which have a direct impact on its fitness for use.

TABLE OF CONTENTS

Section 1:   Quality System

1.1 Establishment

A structured system should be established and implemented to carry out the stated policies and objectives of the organization. This system should provide for implementing those sections and subsections of this reference manual that are applicable to the industry and organization; and should emphasize preventative actions that avoid the occurrence of problems, while maintaining the ability to respond to and correct problems as they occur.

Care should be exercised in establishing the system to maintain a balance between guidance and requirements. Although actions which are essential should be established as requirements, guidance allows greater flexibility in responding to changing situations and provides for greater creativity in fulfilling assigned responsibilities. Guidance is also more readily accepted and supported by the people who must implement the system.

1.2 Updating

An effective quality system should be adapted continually to changes brought about by new or better products/services, customer demands, and changing personnel. Management should not permit complacency regarding the quality system as it will inevitably lead to a system which ceases to meet the needs brought about by continuing change. Its use should not stifle the opportunity for adopting improvements in quality systems as they are developed.

As time goes by, all systems and processes tend to become steadily more complicated without necessarily becoming more effective. As such, simplification should be a prime consideration when updating the system.

1.3 System Formalization

The policies, organization, and administrative systems should be documented in accessible form such as a policy and programs manual (or the uppermost manual in a tiered system of manuals). Care should be taken to enhance the stability of this documentation such as by avoiding unnecessary detail. This information should be accessible by all personnel within the organization. The documents should be revised to reflect policy, organization, and system changes and periodically be reviewed and updated.

Section 2:   Quality Focus

2.1 General

All members of the organization must be responsible for ensuring that they and the organization provide acceptable items and services to their internal and external customers, and should participate in establishing the elements of quality focus. The top member of the organization should be responsible for ensuring that the elements of quality focus are established, documented, correct, appropriate, and implemented; and that the objectives are obtained. The top member of the organization must be responsible for assuring that the organization provides acceptable services and that the organization's customers are satisfied.

2.2 Organization

Clear lines of authority should be established to administer and implement all aspects of the organization. Assignment of responsibility and authority must be balanced and measurable, and the accountability for quality should be assigned from the top member to every individual in the organization. (See "Defining the Organization," Chapter 6, page 47.)

2.3 Long-Term Focus

Brief statements which encapsulate the mission and distinctive competency of the organization should be developed. These will provide the ultimate focus for all activities within the organization. In addition, a set of long-term (one might say eternal) objectives should be established which are brief and address major aspects of the organization including social responsibility, excellence, image, individual and company growth, and innovation. These long-term objectives should be chosen based on their importance in assuring the reality of the mission and distinctive competency statements, and in establishing a work environment reflecting the work ethics and morals of the organization's members. All other aspects of the quality system should be based on the pursuit of these long-term objectives. (See "Establishing Ultimate Focus," Chapter 5, page 42.)

2.4 Quality Policy

Policies provide a broad guide to action, a statement of principle. Policies should be developed for the organization to describe management's specific commitments with respect to the quality of the products/services it provides, product/services it uses, and activities performed. Quality must be the guiding principle, and not be subservient to any other business consideration such as schedules and costs. The policies should specify an organized approach for carrying out those commitments and should address itself to all major quality requirements including safety, liability, adherence to legal requirements, and its specified purpose. Policies should be understood and approved by the chief executive officer for company-wide policies or by subordinate officers for specialized policies, as applicable. Quality system audits and management reviews (Subsections 3.2 and 3.3) should be conducted to assure adherence to the policies.

2.5 Strategic Planning

Once each year, a plan for operation, improvement, and growth should be developed. This plan should address the following three to five years with primary emphasis on the next twelve to eighteen months. (See "Strategic Planning," Chapter 6, page 56.)

Objectives

Short-term objectives (1–5 years) should be established and include objectives for the quality improvement, quality cost management, enhancement of tools and skills, product/service development and modification, and market consolidation/expansion. Objectives should also be established to correct major weaknesses identified during audits and management reviews, and in the areas where major recurring problems exist. These objectives should complement and support the long-term focus of the organization (Subsection 2.3), and be periodically reviewed and updated to reflect changing situations.

Plans

The short-term objectives should be converted into an overall action plan which outlines the methods of accomplishment, primary responsibilities, and realistic fulfillment dates that cause the organization to stretch. Subordinate detailed plans may be developed for major units of the organization or for major objectives. Plans may address key intermediate milestones and result in modification of the overall action plan or objectives. Plans should be periodically reviewed and updated to reflect changing situations.

Goal Formalization

The objectives and plans should be formally documented and presented to, and understood by, the organization. Intermediate milestones and subordinate plans may be formally and periodically documented and provided to affected segments of the organization.

Section 3: Quality Assessment (Assurance)

3.1 General

Mechanisms should be established to assess the current status of quality reflected by the organization; in other words, to provide assurance of quality. These mechanisms should provide for the systematic collection/ generation of information which will assure the recipient that all is well or provide an early warning that all is not well. To be effective, these mechanisms must be designed to avoid extensive personal involvement by those who need the information, and must provide objective informa- tion which is free from the bias of vested interest. The predominant forms for providing quality assurance are:

- Quality audits and management reviews,
- Summarized reports, sometimes referred to as executive reports, status reports, or performance reports.

3.2 Quality Audits

To assure systematic control and effectiveness, quality audits should be made by qualified individual(s) of the organization who are independent of the organizational unit being audited, or by a qualified third party. Audits should be planned, performed, and documented in accordance with documented procedures. These audits should focus on the more important or critical activities (frequently referred to as performance based auditing), verify conformance, evaluate effectiveness, and identify beneficial changes in the quality system. The results should be documented and reported to each level of management which is responsible for the areas audited. Audits may include assessment of the following:

1. Implementation and effectiveness of the policies, procedures, and instructions,
2. Compliance with customer requirements and product/service specifications,
3. Adequacy of quality ensurance measures,
4. Linkage between long-term focus and implementing aspects of the quality system,
5. Implementation and adequacy of the planning, quality assessment, and quality improvment aspects of the quality system.

There are three basic approaches to auditing quality. One or more of these approaches should be used, or they may be combined to form a hybrid.

- Quality System or Quality Program Audits focus on individual programs or on the system as a whole
- Process Audits focus on the effectiveness of the quality system by examining individual processes or grouping of processes
- Product Audits focus on the effectiveness of the quality system by examining products/services after release to the next customer.

3.3 Management Review

The quality system may be reviewed at least annually by a management team to ensure its continuing suitability and effectiveness. This review should not be confused with the quality audits which primarily focus on implementation and conformance, and are performed by independent qualified auditors. Results of management reviews should be provided to every management level.

3.4 Summarized Reports

The bulk of quality assurance is derived from sources of detailed information which is generated during production activities. When summarized and converted into suitable form, this information enables managers to be informed as to quality performance without extensive direct involvement in the production activities. Important aspects of summarized reports include:

1. "Performance measures," or characteristics being reported—selection of the information to be included in the summarized reports is the most critical decision in setting up and modifying a reporting system. Omitting a vital subject deprives management of vital information. Adding unneeded subjects dilutes the attention given to important matters.
2. Comparison to standards—reports on performance have little meaning unless they provide a comparison to some standard which enable managers to judge whether action is needed. This may be accomplished by comparison to competition, historic performance, current goals, technical requirement, etc.
3. Form and content—to be effective, summarized reports must be easily interpreteed by, and meaningful to their target audience. This may be accomplished by
   • Designing reports to meet the needs of their specific audience
   • Including some interpretation to assist managers in assimilating significant facts
   • Identifying opportunities, alarm signals, and progress in pursuing previous opportunities and on implementing corrective action associated with previous alarms.
4. Renewal—reports must consistently provide useful information which is routinely used. Information and reports which are not used should be eliminated or returned to the useful/used category. ("Just routine" can be a clear sign of waste.)

Mechanisms should be established to ensurre that summarized reports are generated and provided in a planned and systematic manner. These mechanisms should also be subject to periodic quality audits to assure the validity of the summary information and assess the effectiveness of the mechanisms.

Quality Monitoring Reports and Project Monitoring Reports are two principle examples of summarized reports, and are discussed below.

3.5 Quality Monitoring

Key indicators of product/service quality and business performance should be identified and mechanisms established to collect, summarize, and statistically evaluate these "Quality Measures." The following are examples of categories and indicators: control of quality including adherence to requirements, perceived customer satisfaction, and scrap and rework levels; growth of the organization's people, including participation and training; service effectivity; quality improvement efforts; quality costs; financial profiles such as return on investment, research and development costs, and profit/loss.

Results of the statistical evaluations of the performance data should, virtually always, be addressed when Quality Measures are reported to management. For example, the report should clearly indicate whether or not apparent changes are real or reflect normal variation in the performance of the system/process. In addition, Quality Measure performance data should be reported in a historical format such as addressing the previous six to twenty-four months.

The mechanisms may provide for a tiered structure of the Quality Measures so that each manager may monitor his/her individual organization. In such a tiered structure, the measures at one level predominantly results from a combination/distillation of the measures at the lower levels; but generally does not reflect all lower level measures.

To each person in the organization, their own set of measures provides feedback as to their performance in producing products/services and in controlling the capability, variability, and level of the production processes. At the top of the organization, the measures represent overall quality trends and performance in specific critical or troublesome areas. (See "Performance Monitoring Systems," Chapter 6, page 54.)

3.6 Project Monitoring

Mechanisms should be established to ensure that management is periodically appraised of the status of important projects, activities, and progress toward goals within their areas of responsibility, or which may impact their areas of responsibility. This may be accomplished by use of the status (or activity) report where each person produces a brief report on the status of projects, activities, and progress toward goals which they and their immediate manager consider important.

The mechanisms may also provide for a tiered structure of status reports. In such a structure, the reports produced by managers are primarily combinations and summarizations of items reported by their employees. At the top of the organization, the report provides summarized information on the status of critical projects, significant activities, and

Don't miss the forest because of the trees. Remember, if you are reviewing the entire reference manual, your objective must be to obtain an understanding of the overall management system and the relationships between the individual elements.

progress toward organizational goals. (See "Performance Monitoring Systems", Chapter 6, page 54.)

Section 4: Quality Ensurance

4.1 General

Although an organization may see itself as only providing products and services to its customers, it also procures and produces products and services for use within the organization; and the quality of its outgoing products and services depends on the quality of the products and services used by the organization. The quality of these products and services is ensured through the development and uniform application of controls which are consistent with quality principles, management policies, and the organization's objectives. The quality of these outgoing products and services is also ensured through managements' consistent focus on quality and implementation of the controls, and through systematic programs for improvement. General categories of quality assurance are addressed in the remainder of this section and in Sections 5–13.

4.2 Instructions and Procedures

Activities should be prescribed in and controlled by instructions, procedures, and guides to the extent that standardization or regulation of the activity is (1) necessitated by process complexity or repetition, or (2) important to the control, improvement, and assurance of product/service qualities. These documents should describe the criteria for verifying satisfactory performance and compliance to requirements. Drafts of these documents should be reviewed by selected individuals who are responsible for issuance, will implement the documents, and will interface with its operation (see Subsection 4.3 for additional guidance). These documents should be immediately accessible to the persons performing the activity. Detailed instructions should not be included in the policy and programs manual (or the uppermost manual of the organization) as this would adversely impact the stability of this manual and violate the guidance provided in Subsection 1.3. (See "Establishing The Procedure System," Chapter, 6, page 57.)

4.3 Document Control

Documents which control activities, stipulate product or service requirements, or provide for collection of information (e.g., forms) are vital aspects of the management system. Methods should be established to control the issuance, distribution, and revision of these documents. These methods should ensure that:

- The documents are reviewed for correctness and adequacy and approved by authorized individuals,
- The documents are released and available in the areas where the work is to be performed,
- The documents are revised as appropriate to ensure adequacy, clarity, completeness, and correctness,
- Obsolete documents are recalled and/or destroyed.

These methods should provide for interim changes and temporary deviations from the prescribed requirements, and may provide for the identification of specific lines of text which have been revised.

4.4 Inspection

Products and services should be determined to meet applicable requirements and be otherwise acceptable before being released for use by another member of the organization or the customer. Principal characteristics of the products and services which are produced or processed should be identified. The products or services should be inspected to ensure that these characteristics are of requisite quality. This inspection may be accomplished on the completed product/service or while in the production process. The results of this determination and the person making the determination should be clearly identifiable. The following two forms of verification may be used to ensure acceptance, as appropriate:

1. 100 percent inspection of product and service characteristics,
2. Inspection or testing of randomly selected products/services or of randomly selected product/service characteristics based on predetermined statistical basis.

Note: The importance of inspection reduces as the process variability is reduced.

4.5 Control of Equipment

Equipment and measurement processes should be maintained in a state of control which ensures that acceptable products and services are being produced, and the verification results are correct and accurate. Measurement devices should be of the type, range, accuracy, and precision appropriate to the characteristic being measured. A maintenance and calibration system should be established so that equipment can be calibrated, adjusted, repaired, or replaced before becoming inaccurate or imprecise. Where appropriate, calibration should be against certified standards having a known valid relationship to national standards.

4.6 Product and Service Unit Control

The inspection and acceptance status of each unit of product and service should be readily identifiable. Such identification may take the form of stamps or notations on work records or inspection and test records that accompany the unit. For example, the signature on a letter or report attests to the completion of inspection and acceptability of the document.

4.7 Breakthrough Management

The quality improvement effort should include a systematically planned and budgeted (proactive) activity. This activity should include all managers in the organization in a tiered fashion with each manager identifying, planning, and budgeting for improvement efforts in their area of responsibility. Overall management (steering) of this activity should be the responsibility of the top member of the organization, and may be performed by a management team which is led by the top member of the organization.

1. There should be an ongoing effort to identify changing market needs, problems/concerns, cost reduction opportunities, and variability in the quality characteristics of products, services, and activities which produce the products and services. The items with the greatest impact on the customer and the organization should be first priority as subjects for the quality improvement program.
2. There should be an ongoing effort to identify characteristics of products, services, and activities for which improvement would have major benefits to the customer, organization, or members of the organization (do not overlook those which are consistently considered acceptable). These characteristics should be selected as secondary subjects for the quality improvement program.
3. Benchmarking may be used to identify and incorporate proven practices that lead to superior performance. Benchmarking is a process for investigating and evaluating superior performance at other organizations and in other industries, and for transferring their strengths into your organization and products/services.

The Quality and Project Monitoring Systems (see Subsections 3.5 and 3.6, pages 93 and 94) provide major sources of information on which these improvement projects are identified, prioritized, and monitored. Strategic Planning (see Subsection 2.5, page 90) provides the planning tool for formally establishing the goals, allocating responsibilities, and defining the overall action plans and milestones.

4.8 Statistical Methods

Statistics is the mathematics of the collection, organization, and interpretation of numerical data. The correct application of statistical methods provides the most accurate information for decision making at the least cost. (For example, the numerous situations and problems faced in life are unequal in importance. Identifying the vital few among the trivial many is an essential aspect of management. A simple method based on the Pareto Principle allows quick determination of the vital few.) All members of the organization (especially management) should have a working understanding of rudimentary statistics and statistical methods.

Statistical methods may be applied whenever numerical data exists or non-numerical data may be grouped or categorized. Statistical methods should be used any place an assessment of a system or process is needed (this includes most problems of management). Some areas where use of statistical methods may be most beneficial include:

- Market analysis
- Process control and capability studies

- Quality improvement investigations
- Development and implementation of inspection plans
- Performance assessment and prediction (including financial)
- Problem analysis
- Preventative maintenance.

Section 5:   Human Resource Management

5.1 General

The most important resource in all organizations is that of it's people. Management should provide a work environment that fosters excellence and a stable/secure work relationship, and should assure that the combinations of materials, equipment, instructions, rewards/recognition, and supervision facilitate continual improvement of each person's performance and growth.

5.2 Functional Definition

Built upon involvement and participation leading to commitment, there should be a mutually developed, agreed upon, and documented understanding between each person and his/her supervisor regarding the individual's functions, responsibilities, and authority. Particular attention should be placed on identifying those aspects of the quality system where the individual will affect the perceived and actual quality of the product/service. This documented understanding should maintain a balance between generality and specificity. Excessive specificity tends to stifle innovation and restrict career growth and spontaneous involvement in the improvement of closely related functions; excessive generality tends to cause conflicts within the organization over functional responsibilities and authorities.

5.3 Indoctrination and Training

Everyone in the organization should be trained or qualified to perform their function in a proficient professional manner, receive periodic training to enhance their competence and performance, and experience cross-training to broaden their understanding of other aspects of the

organization. Everyone in the organization should be aware of the relationship between their function and

- The product/service being provided to the customer, and
- The company objectives and goals.

There should be mechanisms in place to ensure that each person has received suitable training. (Team participation is an excellent method for people to share and broaden their knowledge and experience.)

5.4 Career Planning and Development

Individual performance should be regularly reviewed and discussed with the individual for the purpose of employee growth and development. (The more an individual improves, the more their work will improve and the more valuable they will be to the organization, regardless of whether or not their improvement is work related.) As part of the review process, a development plan should be established which includes specifics regarding objectives, schedules for accomplishment, and organizational resources which will be allocated toward attaining the objectives.

Each person should be encouraged to pursue his/her professional interests within the context of the organization's objectives, expand the scope of their responsibilities, and develop better ways of accomplishing their responsibilities. Excellence requires ongoing, immediate, and constructive recognition of performance with any periodic assessment being an overview of previously discussed accomplishments. Weekly or monthly activity/accomplishment reports may be utilized to ensure specificity during the periodic performance evaluation. Individuals and teams should receive recognition for superior performance and the pursuit of excellence.

5.5 Participation and Involvement

People tend to take more pride and interest when they are allowed to make meaningful contributions to their work and influence decisions made about their work. Mechanisms should be built into the quality system which encourage and provide for broad participation in the organization's efforts and decision making processes. Examples of areas for participation include, planning, establishing policies and procedures, resolving problems, controlling processes, and improving processes, products, and services.

5.6 Motivation

Management should periodically assess and, as appropriate, alter factors which influence motivation. (Identifying and understanding factors in an organization which influence motivation requires substantial study, and is essential to the success of a quality system.) Career development, involvement in the quality focus activities (Section 2), team participation, recognition and reward for accomplishment, and information bulletins are some factors which influence motivation. (For more on motivation, refer to the *Quality Control Handbook* by J. M. Juran.)

5.7 Structured Teams

Structured teams are teams which have been formally established, have a trained team leader, and comprised of members trained in team problem-solving processes and techniques. Management should establish one or more programs for overall management of structured teams, encourage participation on such teams, and provide time and resources for team activities. Structured team programs should provide for recognition of those people who participate on team activities, and rewards for team accomplishments. (See "Structured Teams" in Chapter 6, page 51.) The programs may focus on one or more of the following:

- Improving production, administration, and management processes
- Creating a Total Quality culture
- Improving product/service design
- Creating new products, services, and processes
- Managing/coordinating major projects
- Investigating problems

5.8 Experience Enhancement

Job rotation throughout the organization may be used to promote identification of the organization as a team, broaden experience base, and promote a stable and secure work relationship. (One form of job rotation is when people with similar duties who report to the same manager/supervisor exchange responsibilities. Another form is when people are placed in a multiyear rotation training program and are assigned to open positions for six month intervals.)

Important aspects of a job rotation program include:

- Rotation should be available for those who are interested and not a mandatory activity.
- Each individual's qualifications should be considered when selecting a rotation job. The job selected should enhance the prospects of growth for the individual and organization, yet it should not be beyond the person's short-term ability to grasp and master.

Participation on structured teams is also an excellent tool for enhancing team environment, experience base, and work relationships.

Section 6:   Product and Service Development and Change

6.1 Development Plan

A plan should be established for development of each new product and service or substantial modification in an existing product/service. The plan should address the identification of customer needs, marketability, and validation and audit methods for appropriate stages of development and implementation. The plan should also address resource allocation and development milestones.

6.2 Product/Service Specification

A specification should be established to define each new product and service, and revised when substantial modifications are made to an existing product/service. The specification should address the purpose for and scope of the product/service, the applicable customer and regulatory requirements, and the methods to be used to control and assess performance of the product/service. It should identify the primary characteristics of the product/service, and may specify acceptance criteria for these characteristics. The specification may specify the measurement devices and systems, and may state accuracy and precision requirements for the measurement devices and systems. In addition to the customer, the following functions should be involved in developing the specification: marketing, design, production, distribution, sales, delivery, and field services.

6.3 Product/Service Validation

New and substantially modified products and services should undergo validation to ensure that the product/service and processes for production and administration (including the marketing strategy) are fully developed, and that the product/service meets the needs of the customers and complies with the policies and objectives of the organization. Validation should be completed prior to marketing the product/service. Management may attest to the acceptability of the product/service and authorize marketing in writing. Planning should provide for appropriate time-phased milestones at which key events during the development or modification process are monitored or verified to be acceptable. Validation may be accomplished through one or more of the following.

FORMAL REVIEW.   A timely independent review of specifications and technical documents related to product/service development, safety, and performance should be performed and documented. Representatives of customers and each function concerned with the product/service should participate in the review to assure that all necessary information and requirements are clear and specified. Examples of elements of a review may include:

1. Specific customer or user quality requirements.
2. Requirements specified by regulatory agencies and appropriate voluntary industry groups.
3. Careful weighing and evaluation of safety requirements.
4. Implementation of the design, including special needs, mechanization, and automation where appropriate.
5. Inspectibility and testability.
6. Clearly defined acceptance criteria.
7. Effectiveness of production and administrative processes.
8. Reliability and maintainability requirements.
9. Ease of problem diagnosis and correction.
10. Impact of environmental requirements.
11. System obsolescence.
12. Historical data relating to personal injury, damage, liability, and environmental degradation.
13. Unintended uses and misuses that can be anticipated and identified.

QUALIFICATION TEST. An actual or simulated delivery and use of product/service or a subcomponent of the product/service. Qualification tests are used to assure that the characteristics of the product/service will be acceptable under anticipated usage conditions. Such tests should be clearly defined, planned, and documented prior to performing the test, results should be documented, and the test should provide for the following, as appropriate:

1. Evaluation of the product/service and its specification in meeting customer needs and exceeding customer expectations,
2. Evaluation of conformance to product/service specification,
3. Controlled testing and a feedback reporting system documenting the results during the qualification test cycle,
4. Posttest reviews of test data to assure that any adverse conditions affecting the required quality of the product/service are corrected,
5. Evaluation of continuity, consistency, timeliness, and efficiency in delivery of the product/service.

FIELD SUPPORT REVIEW. A review to determine whether field support is adequate and prepared for the new or modified product/service.

This should be a planned and documented activity, and should consider the principal aspects of field support such as:

1. Availability and adequacy of instructions, procedures, and guides,
2. Training of personnel,
3. Satisfactory completion of qualification tests,
4. Need for or adequacy of customer service organization.
5. Feedback of problems discovered in the field.

6.4 Revalidation

Periodic reassessment of products/services may be performed to ensure that they continue to meet the needs of the customers, comply with the organization's policies, and conform to the applicable product/service specification. Revalidation should also be used to identify potential improvements in the product/service and processes for production and administration. Revalidation should be a planned and documented activity, and should include consideration of actual field experience, impact of minor modifications in the product/service and processes, impact of personnel changes, adequacy of procedures/instructions/guides, and proposed modifications.

6.5 Change Control

Modifications in product/service and processes for production and administration, and changes to the product/service specification should be controlled to ensure that the changes are appropriate, do not violate established requirements or organizational policies, do not adversely impact accomplishment of organizational objectives, and will not be or be perceived as a reduction in quality by the customer.

Section 7: Marketing

7.1 Market Analysis

The existence of every organization depends on its ability to market its products and services. This includes identifying new products and services, new markets for current products and services, and current products and services which are no longer marketable. It also includes identifying: What the consumers think of current products/services?, Why previous customers no longer use the organization's products/services?, and Why potential customers are not using the organization's products/services? As such, market analysis is an essential aspect of product/service development and customer satisfaction management. A market analysis program may include the following elements:

1. Customer surveys may be used to ensure that customer requirements, needs, and expectations are identified and satisfied; identify and evaluate the competition and the performance of their products and services; and identify additional products and services which could be supplied to current customers.
2. Safety and environmental legislation and related national and international standards and codes may be reviewed and compared against product/service specifications and contracts to ensure compliance.
3. Market surveys may be used to identify additional markets for current products and services, and to identify potential market needs which could be met through the addition of new product and service lines.
4. Competition analysis may be used to evaluate the competition and the performance and cost of their products and services. This may

be done by an independent firm which specializes in collecting this commercially sensitive information.

7.2 Contract Management

Contracts establish the actual or perceived basis upon which products and services are provided, and may be in written, oral, or implied form. Although the use of oral and implied contracts should be minimized, these contracts form the basis for most services provided directly to the consumer. To minimize misunderstandings which are prone to occur from oral and implied contracts, the customer should be provided information explaining product/service scope, process, and compensation. This information should be clear, brief, and readily available to the customer, and should address warranties and customer feedback interfaces. Written contracts should be established through three phases.

PHASE ONE   The first phase is one of definition and understanding. Prior to bidding, the organization should:

1. Establish a liaison with the potential customer.
2. Understand the needs of the customer, such as delivery and reporting schedules and details of applicable specifications and drawings.
3. Be aware of essential contract requirements including regulations, standards, codes of practice, and regulatory interfaces which the customer will expect to be implemented within the scope of the contract.
4. Ensure that the needed products and services can be provided within the scope of an existing product/service specification or ensure that a new specification or the production processes are developed.
5. Ensure that resources will be available to execute the contract, and that affected internal and supplier functions are aware of and agree with the proposed contract requirements.

PHASE TWO   The second phase is one of verification. Prior to approval, the contract should be reviewed by affected functions and may be reviewed by an independent reviewer to verify that:

1. Phase One items 2 and 3 are appropriately and correctly included in the contract;

2. The product/service can be effectively provided within the terms of the contract, e.g., that the product/service specification has been approved or the processes have been established;
3. Resources are available to execute the contract and the affected suppliers are prepared to fulfill their aspects of the contract.

PHASE THREE   The third phase is implementation. After contract approval, the following should be performed.

1. All affected functions should be made aware of contract details which affect their activities, and assurances may be obtained from the functions that the contract will be executed as written.
2. Execution of the contract should be monitored to assure compliance.
3. Contracts should be promptly revised as necessary to ensure accuracy, understanding, and compliance.

7.3 Selling Quality

A comprehensive and effective quality system may be utilized as a marketing advantage, through such activities as customer involvement teams, customer awareness programs (e.g., in-house tours, employee visits to the customer), and advertising. However, the marketing organization should ensure that the organization is aware of the liability risks and financial implications of offering exaggerated claims regarding the quality/performance of the products/services.

Section 8:   Procurement

8.1 General

Procurement activities should be planned and documented to assure effectiveness of the procurement process. The documented process should include identification of procurement methods and organizational responsibilities, and establishment of procurement quality requirements. The procurement process should reflect a graded approach based on the importance of the specific procured product or service to the ultimate product/service and its impact on the organization, customer, and society. Procurements should not be based on the price tag; but should be based on the total (life cycle) cost which includes the costs of maintenance, handling and reworking nonconformances, disposal, potential production delays, and warranty when nonconformances reach the customer.

8.2 Methods

The documented methods should address planning for procurement, establishment of procurement specifications, selection and qualification of suppliers, bid evaluation (where appropriate), procurement authorization, verification of product and service quality, control of conditions adverse to quality and corrective action, acceptance of product and service, evaluation of supplier performance, and control of records.

8.3 Planning

The plan should be based on a prioritized list of needs and be revised when necessary to reflect changing needs. The plan should include consideration for routine and crisis procurements and for maintenance and replacement of major assets.

8.4 Supplier Quality System Control

To the degree consistent with quality policy and a graded approach to procurement, the purchaser should place appropriate quality system requirements on the supplier. For important and moderate to high impact procurements, a determination should be made as to the elements of the quality system that are applicable to the supplier in assuring quality; and these elements should be imposed on the supplier through the contract. The contract should also include provisions for right of access, document submittals, and warranty.

8.5 Quality Assessment

Procedures for determining acceptability of any purchase should be based on the graded approach and instituted by one or a combination of the following:

SUPPLIER QUALITY EVIDENCE. Supplier generated quality information may be used, where practical, to avoid unnecessary or redundant inspections and tests when all of the following conditions exist:

1. The type of information is defined and specified by the purchaser.
2. The information is a statement of fact or condition based on quantitative or qualitative observations.
3. The information can be verified by the purchaser.
4. A periodic verification program is established.

INSPECTION AT SOURCE. Source quality control by the purchaser involves the performance of specific quality checks at the supplier's facility or surveillance of the suppliers' quality control processes, acceptance procedures, and quality records. Where source control is utilized, it should include a review of applicable quality system elements. Source quality control is applicable when any of the following conditions exists:

1. The cost of establishing and administering such a supplier control plan is commensurate with benefits to be derived, such as better control of quality or more rapid follow-up on nonconforming conditions.
2. The conformance of purchased products and services to requirements cannot be adequately determined upon receipt.
3. Direct shipment is planned for units of purchase from the supplier to the purchaser's customers or sales organization without being routed through the purchaser's verification activity.
4. The cost of duplicating inspection and test activities would be prohibitive.

RECEIPT INSPECTION. For those purchases where verification upon receipt is used, examination should be performed to the degree and extent specified to determine acceptability. Receipt verification should include:

1. Well-maintained records of verification history so that past supplier performance is available. (Supplier performance history should be statistically evaluated when deciding the characteristics to be inspected and the level of inspection.)
2. Adequate methods for identifying purchased units. Identification should include the test and acceptance status of the units, as applicable, and traceability when required.

3. Identification and verification of units accepted at the source to validate the documentation of testing and acceptance.
4. Supplier's certification of quality.

8.6 Problems/Concerns

Problems/concerns regarding procured products and services should be identified, documented, and resolved, and the problem/concern information fed back to purchasing and suppliers for corrective action.

8.7 Supplier Performance Evaluation

Each supplier's quality performance should be periodically evaluated. A formal rating system will aid in determining effectiveness and planning future quality assurance controls, consistent with the complexity of the product/service and the requirements established in the purchase agreement. Evaluation may include on-site audit of the supplier's quality program. (The importance of on-site audit increases dramatically as the impact of poor supplier quality on the organization increases.)

8.8 Just-in-Time, Sole Source Strategies, and "Internal" Suppliers

As an organization's internal production processes become more effective, the incoming product stocking levels and the variation in incoming product/service becomes more important.

Moving toward a single supplier for critical products/services can substantially reduce internal processing costs which result from the variation in the incoming products/services. Implementation of a Just-In-Time procurement program can substantially reduce the unproductive financial resources wasted in maintaining extensive stocks and multiple suppliers.

However, implementation of these strategies should not be taken lightly, for with the increased benefit comes increased risk. It is crucial that these strategies be instituted in line with the supplier's demonstrated ability to meet the quality requirements, and based on a long-term relationship of trust and loyalty. In a sense, this means developing high quality suppliers through a relationship based on both organizations seeing themselves as pursuing the same objectives and as part of the same organization. (Just-In-Time programs tend to be the natural outcome of a good quality management process; it is *not* a starting point. Just-In-Time efforts should never be attempted until the internal processes are in statistical control.)

The organization should form a partnership with the suppliers in resolving problems and in developing and improving supplier quality systems and process control techniques. (This becomes very important as Just-In-Time and Sole Source Strategies are initiated.) However, this should be undertaken slowly, starting one supplier at a time until the "internal" supplier program has demonstrated success.

Section 9: Process Management

9.1 General

Processes (set of interrelated activities, methods, machines, or tools used to attain the qualities desired for products or services) should be controlled; this is a substantially more effective method for ensuring acceptable quality than that of product/service unit inspection. All activities performed in an organization are part of one or more processes. Many of the processes may be limited to one functional area or organizational unit; although most processes involve more than one functional or organizational unit.

Process management is itself a process which can be subdivided into subprocesses, for example, process definition, control, monitoring, corrective action, and improvement.

9.2 Process Definition

The first step in managing a process is defining the process. Some aspects of process definition are as follows:

1. Establish ownership,
2. Define the boundaries (e.g., inputs and outputs, and involved functions or units of the organization),
3. Define the set and sequence of work activities,
4. Establish the characteristics to be measured;
   • Identify the principal characteristics of the inputs and outputs,
   • Determine in-process characteristics which substantially affect the principal characteristics of the outputs,
5. Establish points of control, as appropriate,
6. Establish measurements which are indicative of process performance,
7. Document the results of this effort, including development of appropriate instructions and acceptance criteria.

9.3 Process Monitoring and Corrective Action

The basic control device is the feedback loop. The principal characteristics of the services and products, and of the processes which influence these product/service characteristics should be used to monitor and control the process. Measuring/monitoring instruments should be developed and used to collect real-time data on these characteristics, and to feed the data back to those people who produce the product/service and those people who are responsible for the process.

In-process inspection should occur soon after the process points which affect the product/service characteristics being examined, and should verify that critical process parameters are being properly maintained. Mechanisms should also be established to obtain prompt feedback from the users, thereby providing additional input for process control. Inspection and user data should also be promptly fed back to the product/service producers and the people responsible for the production process.

Statistical process control techniques should be used to enhance the effectiveness of these functions. Processes which exhibit excessive variation or inappropriate aim (produce unacceptable products and services) or which are inefficient should be analyzed and corrective action instituted to establish an acceptable state of control in the process. Use of statistics in process monitoring also provides data on process capability and performance which substantially improves decisions regarding allocation of resources for improvement (breakthrough).

9.4 Special Processes

Special controls should be established for processes that affect product/service characteristics which may not be easily or economically measured by in-process techniques or inspection, or which may exhibit inferior performance only after use, for example, welding and computer software development. Some of the special controls which may be used are as follows.

PROCESS CAPABILITY VERIFICATION. The ability of the process to inherently produce acceptable characteristics may be studied. Processes which have unacceptable inherent variability or inappropriate centering should be the subject of a process improvement program.

QUALIFICATION TEST. People, procedures, instructions, and equipment may be tested under actual or simulated conditions. Qualification tests are used to assure that important characteristics of the product/service will be acceptable under anticipated and adverse conditions. Such tests should be clearly defined, planned, and documented prior to performing the test, results should be documented, and the test should provide for the following, as appropriate:

1. Evaluation of conformance to customer, organization, and regulatory requirements such as safety and performance,
2. Controlled testing and a feedback reporting system documenting the results during the qualification test cycle,
3. Posttest reviews of test data to assure that any adverse conditions affecting the required quality are corrected.

PERIODIC VERIFICATION.   The continuing capability of special environ-ments, equipment, and individuals to meet specific quality requirements may be periodically verified.

9.5 Process Improvement

Process improvement is the single most effective approach to improving quality and reducing cost. It applies to all processes, including order entry, customer billing, budgeting, manufacturing, design, structured team management and operation, etc. Improvement in work processes should be pursued from two directions:

1. Management Directed—based on management monitoring of per-formance, prioritizing improvement efforts, and initiating process improvement efforts (which is most effectively accomplished by forming a structured team to address each problem area).
2. Self-Initiated—based on individual and group awareness of prob-lems, investigation, and correction proposal (most effectively ac-complished by developing a culture of continuous improvement and through the use of self-directed structured teams, such as Quality Circles).

In each of these approaches, broad use of quality measuring/monitor-ing instruments and reporting systems is essential.

The general principles upon which process improvement should be founded include:

- All processes can be improved to reduce variation in the resultant products/services, production time, and cost; and to increase pro-ductivity, effectiveness, and product/service quality
- At any point in time, roughly twenty percent of the problem sources will be responsible for roughly eighty percent of the problems im-pacting product/service quality, customer satisfaction, productivity, safety, profit, etc. (application of the Pareto Principle)
- Everyone should be involved in process improvement
- Data on quality and productivity, and statistical evaluation of this data are essential in identifying and prioritizing process problems.

Section 10: Records Management

10.1 General

Records should be identified, developed, and maintained to objectively demonstrate effective operation of the quality system, attest to the quality of products and services, and aid in identifying, correcting, and preventing specific problems and concerns.

10.2 Examples of Records

A record may be any document which has been approved and reflects actions taken or to be taken, this includes procedures and instructions, inspection and test results, acceptance authorizations, certifications of quality, cost reports, problem/concern reports, contracts.

10.3 Control

A policy and procedure should be developed which specifies the type and quantity of information to be gathered and retained commensurate with the amount of control needed, the retention schedule, and appropriate legal and contractual requirements. This document should provide general development and issuance requirements as addressed in Subsection 10.4. This document may address record maintenance considerations such as protection from fire, theft, pilferage, electromagnetic degradation, and water damage.

10.4 Content and Use

Each procedure and instruction may provide guidance in the content, handling, and use of specific records generated during the activity being prescribed. Records should:

1. Be current, complete, accurate, legible, and pertinent,
2. Contain information such as date of origination, inspection date, procedures followed,
3. Be identifiable as to product/service unit, and individual and supervisor responsible,
4. Show quantity, type, and severity of discrepancies, as applicable,
5. Be retained according to stated record retention requirements.

Section 11:   Problems/Concerns

11.1 General

Measures should be established to identify, report, and control problems and concerns to ensure that they do not adversely impact the user (internal and external) of the product or service in either an actual or perceived manner. Each person in the organization should be responsible for problem/concern identification. Use of the product or service reflecting the problem/concern should be controlled or precluded until the problem/concern has been evaluated and resolved. Each problem/concern should be documented to ensure that the condition is evaluated, resolved, and reviewed for repetition. Remedial corrective action should be taken to ensure that the immediate condition is corrected. Recurrence control should be instituted for significant and repetitive problems/concerns. Recurrence control is the identification of the underlying cause of the condition and the establishment of corrective actions to eliminate or control the cause. The results of the evaluation and actions taken to resolve the problem/concern and prevent recur-

rence should be reported to persons identifying the problem/concern and to persons affected by the condition and the corrective action.

11.2 Prevention

The most effective approach to handling problems/concerns is to proactively prevent them. This is accomplished through devoted attention to improvement, e.g., in the production, administration, and management processes, abilities of the organization's people, management techniques, supplier capabilities, customer relationships, and all other aspects of operation. However, when a problem/concern does occur, prevention is also accomplished through devoted attention to resolution and recurrence control.

11.3 Problem/Concern Detection

Original indication of a real or potential problem may come from a variety of sources including an inspection or test, customer complaints, a quality audit, quality trend analysis, or external government or consumer agencies. In any case, once a real or potential problem is detected, it should be documented and reported to the appropriate persons for evaluation and correction.

11.4 Responsibility for Corrective Action

Responsibility for corrective action should be specifically assigned. This may include coordination, evaluation, recording and monitoring corrective action efforts to assure that action is taken promptly, and that the action is effective. Specific functions found responsible for conditions which led to the problem/concern should be responsible for determining corrective action and for taking the action. If corrective action is demonstrated to be ineffective, the entire corrective action process should be reinitiated.

11.5 Recurrence Control

In addition to correcting the immediate condition, each problem/concern which has or could have a major impact on the customer or organization should be evaluated and action taken to reduce the risk of recurrence. Merely fixing the immediate symptom of the particular problem does not achieve corrective action. Corrective action should include analysis of specific modes of failure and analysis of the quality system, and result in changes to the quality system or processes. Corrective action procedures should assure prompt remedy determination, incorporation, and verification of effectiveness.

EVALUATING THE NEED FOR RECURRENCE CONTROL.   Since no individual can be expected to recognize all instances in which recurrence control may be both justified and economically feasible, data from all sources should be categorized, and periodically statistically analyzed to identify chronic or recurring problems.

DESIGNATING RESPONSIBILITY.   The responsibility for cause evaluation and corrective action should be assigned to the person who has primary responsibility for the processes which produced the problem.

DETERMINING THE CAUSE.   Determining the relationship between cause and effect is essential to effective recurrence control. The following steps should be taken in determining the cause of a specific problem.

1. Assessing Impact. The extent of the problem should be evaluated in terms of its potential impact on such considerations as avoidable cost, performance, safety, and customer satisfaction.
2. Identifying Variables. Variables affecting the capability of the product/service to meet customer needs and required standards should be identified. The data may be expressed in the form of physical measurements or attributes; for nonphysical characteristics, a scaling or ranking may be useful, e.g., a ranking from 0 to 5.
3. Process/System Evaluation. The processes/systems which control the production of the variables associated with the problem should be evaluated to identify the causes for the problem. This evaluation should include a study of process flow and the use of statistical methods to identify and understand the causes. This evaluation should also result in an understanding of the changes needed to prevent recurrence. (The evaluation should focus on the process/ system causes and avoid personal blame. However, understanding the level of the organization which is responsible for controlling the causes is very helpful in identifying the root causes and needed

changes. The guidance for this determination is as follows: Management is responsible for all causes except when the process is statistically capable of consistently producing acceptable output and the operators:

- Know exactly what is expected,
- Have the tools and are able to determine how well they are conforming to the expectations, and
- Have the tools and are able to adjust the process if it is not conforming.)

INCORPORATING CHANGE. Having identified the root causes for the problem, appropriate controls should be established and their effectiveness verified to assure that affected variables can be maintained within the established limits. These new controls may be included in the list of characteristics regularly verified and reported by the quality system.

11.6 Management Reports

Reports on problems affecting quality should be directed toward management for attention and action. A good reporting system has a hierarchy of reports ranging from hourly through daily, weekly, monthly, and annually, with each type of report having distinct objectives. Each report permits or requires a particular level of management to take action in its own area of responsibility. Another technique to display problems/concerns is the quality trend graph, including comparison of objectives and past performance statistics. This can provide a baseline for recognizing both improvement and deterioration.

Section 12: Financial and Accounting Information Management

12.1 General

The ability to evaluate, project, and guide an organization's performance is only possible when sufficient and reliable information is readily available. Critical questions such as "Why have sales increased and cash flow decreased?," "Why have profit margins dropped on overall or on specific product/service lines?," and "What has caused overhead to increase dramatically when sales are stable?" cannot begin to be addressed without proper financial and accounting information.

As each organization and each unit of an organization is different, the detailed financial and accounting information needs are different. Unless managers have access to the information which fits their unique needs, they will have little or no success in solving the day-to-day problems which they face. These needs fall into three general categories:

- Fulfilling governmental tax authority requirements,
- Demonstrating compliance to contract requirements and financial status to other third parties such as financial institutions, and
- Information for internal use including performance monitoring, strategic planning, and quality improvement investigations.

The first two of these categories are generally considered customary aspects of financial management, and organizations which limit their data collection to these two categories soon discover that they lack the necessary data to solve problems. The last category addresses a substantially larger pool of information and is an important aspect of a Total Quality System. (As with all data used in management decision making, statistical methods should be used in evaluating financial and accounting information; and management reports should address the result of this evaluation.)

12.2 Financial Monitoring System

The accounting and financial information needs throughout the organization should be identified, and systems instituted to collect, analyze, summarize, and report the information (for other system considerations, See subsection 3.4, "Summarized Reports," page 92). This system may be linked to the "Quality Monitoring System" (Subsection 3.5, page 93) to provide one reporting source for upper management and minimize the potential for financial considerations to take precedence over the other quality considerations.

In defining the information needs, consideration should be given to the following:

- What is the structure of the accounting system which best reflects the organization's structure and profit centers? For example, is the class of customer, product/service line, or organizational unit more important in assessing performance and reaching product/service cost decisions?

- Which expense categories are variable with sales or provide reliable information on quality performance such as warranty costs, resource waste, and results of quality improvement efforts? These categories may deserve separate accounts.
- What financial information may be integrated with nonfinancial information to produce effective performance indicators, such as a ratio of material or personnel costs to units of product/service produced?

12.3 Quality Cost Analysis

Quality cost analysis is a financial accounting tool for monitoring and optimizing the effectiveness of the Quality System and for identifying areas where quality improvement is most needed. Another major reason for extending the accounting system to collect quality costs is to bring attention to previously uncontrolled or hidden costs. As generally used, quality cost analysis focuses on the costs attributable to poor quality and are broadly grouped into the following categories:

- Internal Failure Costs—these are costs which would disappear if there were no internal failures, such as scrap, rework, repair, yield losses, and problem control and corrective action efforts.
- External Failure Costs—these are costs which would disappear if there were no external failures, such as warranty charges, returned material, and complaint handling. The estimated cost of lost reputation and customers may also be included in this category. (*Note.* These subjective costs are very difficult to quantify and tend to increase doubt regarding the validity of the whole system when included.)
- Appraisal Costs—these are costs incurred to identify the condition of the product/service during production, such as in-process and final inspection, supplier surveillance/audits, incoming inspection, and products destroyed by destructive tests.
- Prevention Costs—these are costs incurred to reduce the other cost categories, such as quality data acquisition and analysis, quality improvement teams efforts, and new product/service review activities.

(Management must understand when using this tool that it does not report the "true" costs of poor quality because the biggest costs are unknown and unknowable—the cost of an unhappy customer.)

A system for collection, analysis, and reporting of (poor) quality costs should be developed and instituted, where clearly justified (benefits should outweigh the extensive effort associated with development and use of a quality cost system). Development should be based on clearly defined objectives. Quality costs should be periodically reported to management, and be presented with a comparison to standards such as historical numbers or current goals. As quality costs can increase due to such things as organization growth and inflation, the reports should also present the quality costs as ratios to other measures of performance such as sales volume in dollars or units, overall production effort in hours or dollars, profit, etc.

Section 13:   Customer Satisfaction

13.1 General

Customer satisfaction should be the founding principle of the organization and the utmost objective of every person associated with the organization; this applies for internal as well as external customers. This section principally addresses special considerations and controls associated with management of external customer satisfaction; however many of these considerations and controls may also be applied to ensure internal customer satisfaction. Management mechanisms for internal customer satisfaction are addressed throughout this reference manual.

13.2 Interface Specifications

A major element in managing customer satisfaction is the clear and mutual understanding of the:

1. Customer's intended use of and expectations for the product/service,
2. Scope of and compensation for the product/service,
3. Important product/service characteristics, including reliability, timeliness, and safety, and
4. Points of interface between the organization and the customer.

This understanding should be documented and agreed to, and periodically reviewed by the organization and customer where feasible. Much of this understanding may also be in the form of or referenced by a contract. This information is also crucial to the establishment of a meaningful product/service specification as addressed in Section 6.

13.3 Customer Feedback

Well-planned continuing feedback from customers should be obtained to assess product/service quality, including reliability, performance, and overall satisfaction. This information should be analyzed for collective as well as individual meaning; this requires summarizing the data in various ways such as by product/service type, problem/concern classification, length of relationship and use, type of application, etc. The results of this analysis should be reported to management and improvement actions initiated as appropriate. The following are examples of feedback channels which may be used where appropriate:

1. Periodic questionnaires distributed on a random basis,
2. Periodic personal contact by a person perceived to be in authority and knowledgeable yet preferably independent of the production and delivery processes,
3. Customer concern resolution system (see Subsection 13.4),
4. Problem resolution system,
5. Field reports,
6. Marketing performance reports.

13.4 Customer Concerns

Customers should be urged to provide notification of their dissatisfaction. Mechanisms should be established to ensure that particular attention is focused on each concern raised by a customer and that the concern is promptly resolved. The mechanisms may also provide for follow-up verification of customer satisfaction by a person perceived to have authority in the organization, yet not directly responsible for the condition which caused the concern. The customer concern control mechanism may be an integrated but unique element of the quality problem/concern resolution system (Section 11).

Appendixes: Examples

Once again, it is important to remember that the procedure and process examples provided in this part of the guide are provided as examples of how to implement and combine the system guidance provided in the "Quality System Desk Reference." These examples should not be applied as written to any organization, but may be revised to reflect the organization's new focus, strengths, weaknesses, and current related processes.

Team Problem-Solving Techniques

This appendix provides examples of some of the more popular and simple problem-solving techniques used by structured teams. Most of these techniques are also quite valuable when used by individuals during the decision making process (in many companies, management is required to use some of these techniques such as statistical based charting before making major decisions).

Brainstorming

A divergent thinking process by which a large number of ideas can be generated. This process is used to stimulate creativity by building upon the thoughts of others. Rules include: everyone participates, no criticism or comments, people can pass. Procedure:

- Each person speaks in turn (round robin)
- One member records ideas exactly as reported for everyone to see

- No one censors or interrupts anyone else
- Group aims for 35–100 ideas and unrelated ideas are accepted
- Hitchhiking on another's ideas is encouraged; modify, magnify, reverse, combine, substitute, or reduce

Nominal Group Technique (NGT)

A divergent-convergent technique for creating a large number of ideas and effectively ranking their importance. This technique promotes balanced participation. Rules include: everyone participates and ideas are written.

Procedure:

- Each person silently writes down at least three ideas, one idea to a card, maximum phrasing of 5 words to an idea
- All ideas are listed on board or flip charts after shuffling the cards, and each idea is given a sequential number
- Ideas are discussed, clarified, combined, added, deleted, and modified; prolonged argumentation is avoided
- Each person silently selects five best ideas and ranks them 1–5 with 5 being the best; outside-in ranking process used (select best, worst, next best, etc.) and results are put on cards (one card per idea)
- All cards are collected, rankings are transferred to board, and totals are calculated for each idea; highest score indicates group consensus
- If no clear-cut consensus, low ranked ideas are execluded and a second round of ranking is conducted

Multivoting

A convergent technique for reducing the number of ideas down to a manageable size. Rules include: number of permitted votes cut in half with each round, and the objective is to reduce the number of ideas not arrive at one choice. Procedure:

- Each person votes for as many items as desired, but only one vote per item
- Items with the lowest half of the votes are removed from consideration
- Repeat voting with each person only allowed to vote for approximately half of the remaining items

- Pick top half of items with the highest votes
- Repeat voting-reduction process until the desired number of items remain

Pareto Analysis

A technique for making the "vital few" and "trivial many" obvious to an observer. Examples of the use of this technique include identifying the most significant problems from a list of problems, the most significant causes from a list of causes, most important customers from the customer base, etc. Procedure:

- Record all items in a list, one item per line; consider grouping clearly minor items under one item named "other"
- Select a unit of measurement which applies to all items such as frequency of occurrence, dollars, hours, units
- In the column next to the item list, record each item's measure (e.g., frequency of occurrence); at the end of this column, record the total of all measures
- In the next column, record each items percent-of-total
- Reorder the list by percent-of-total with the largest number at the top
- In the next column, place the cumulative percent-of-total
- The "vital few" may be defined as those items which account for 75 to 80 percent of the total measure and will be readily apparent from the cumulative percent column

Graphs, Histograms, and Charts

Techniques for visually presenting a clear picture of the facts.

- A *Line graph* consists of a horizontal line labeled with the nonvariable data (e.g., date) and a vertical line labeled with the variable date (e.g., number of defects). These two lines generally are drawn such that the left of the horizontal meets the bottom of the vertical line. The data is plotted on the graph and a line is drawn connecting the data points. A line graph may have more than one set of data plotted (more than one line).
- *Column graphs* are constructed like a line graph but a wide line (column) is drawn up to each point from the horizontal line with a

gap between each column (instead of the points being connect). Also, all columns are of the same width.

- A *Bar graph* is a line graph on its side (the wide line is now horizontal instead of vertical).
- Although bar and column graphs are quite similar, they appear differently to the human eye. Column graphs should be used when the relationship between the data is to be visually enhanced. Bar graphs should be used when it is desired to draw attention to each individual bar.
- *Histograms* are similar to column graphs and focus on the frequency of occurrence in sequential order. In a column graph, there are gaps between the data point columns; whereas, in a histogram, a single vertical line provides the boundary between the cells, giving a stairstep appearance. Histograms are used to focus attention on the historical trend.
- A *Pie chart* is a circle divided into pie shaped wedges with each wedge representing a segment of the data. Pie charts are useful for showing 100% of something and emphasizing relationships between the individual segments and the whole. For example, product line contribution to total profit may be depicted on a pie chart where each piece of the pie represents a product line's percentage of total profit.
- *Pert* and *gant charts* are used to plan, schedule, and control complex projects. They identify the variables and relationships which affect schedule, timing, and costs.
- *Process Flow Charts* are used to depict the sequence of steps in a process. In constructing a process flow chart, different shape blocks are frequently used to indicate different types of activities (such as a diamond for decision). The sequence of the process steps is depicted by lines connecting the blocks and the process flow is depicted by an arrow at the end of each line.

Procedure:

- Collect information
- Determine which visual presentations will best reveal the message
- Develop the graphs, histograms, charts
- Analyze the message from the visual presentations
- For presentation to others, refine the visual presentation to ensure that the message will be clear at a glance (keep it simple and uncluttered)

Statistical-Based Charts

A broad category of techniques which combine the benefits of clear visual presentation for graphs, histograms, and charts discussed above with the power of statistics.

Control charts (possibly the most widely used statistical-based charts) are used to monitor a process and identify when process adjustments are appropriate (identify problems before or as they occur). Control charts are similar to graphs except horizontal lines are added to identify the "control limits." Also, each control chart data point represents the average of a process sample and is plotted at the time of sampling. Use of a control chart provides immediate information as to the probability that a shift has actually occurred in the process; and thus substantially reduces the contribution to process variation from errors in decision. The following is an example of a control chart.

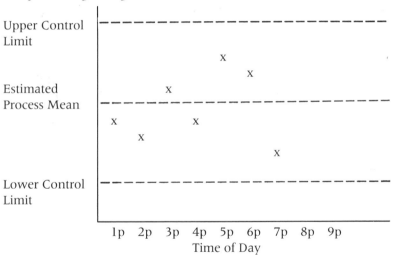

Control charts are designed to be used to monitor a process which is known to be under control (variation in data is random and within acceptance limits). They are also relatively simple to construct and use because the limits are obtained from simple calculations and tables and the rules for interpretation are few and simple. For example, the above chart depicts normal variation in the process as revealed by no points outside the control limits, no long sequence of up or down runs, and less than six sequential points on one side of the process mean.

Although the control chart technique is designed for use in a real-time situation and on a process which is under control, the technique can also be used after the fact to aid management in decision making. For example,

- One key quality indicator (on time delivery) has been close to 92% for the last two months but has dropped to 90%. Should management initiate an investigation as to the cause of this drop or use its resources in other vital areas? The answer to this question would be clear if management received the information in the form of a control chart.
- Your best client has complained about receiving an outrageous bill for services rendered and you have committed to investigate the situation, adjust the bill, and take action to prevent it from recurring. You have just received the report on the investigation. It indicates that the overall job consisted of six projects and that the five routine projects were performed by a junior partner who takes an excessive amount of time on projects; and it recommends cutting the bill in half and reassigning the junior partner to other duties for which his abilities are better suited. Is the reported cause correct and is it the root of the problem? Is the recommendation appropriate? Is the problem really linked to the productivity of one person or is there a department wide (systematic) problem? If the report contained a control chart on each project category with the average project times depicted for each person in the department, you would probably have the answer to these questions. *Note:* Other statistical techniques would have been more appropriate in the above situation. However, knowledgeable (although imprecise) use of simple control chart techniques will generally produce adequate information for decision making; and, over the long haul, significantly more decisions will be correct than if no statistical techniques are used.

Cause and Effect Analysis (also known as Ishikawa Diagram or Fishbone Diagram)

A technique to help visualize a problem and visually represent probable causes in specific categories. Rules include: The problem is a measurable product or process, and Anything that may result in the effect is considered a potential cause. Procedure:

- Problem is stated in the "effect" box on the right of the board and a horizontal line drawn from the left to the box
- Major categories for the possible causes are determined and recorded in "cause category" boxes which are above and below the horizontal line; a diagonal line is drawn from each "cause category" box to the horizontal line; materials, manpower, methods, and machines make up one commonly used set of cause categories
- Brainstorming and fact finding are used to generate possible causes, each potential cause is discussed to clarify, combine, and identify appropriate cause category(ies); diagonal lines are made to the cause category lines and one potential cause recorded on each line under the appropriate category
- Convergent thinking techniques such as NGT or multivoting are used to focus on the most probable causes
- Further searching for root causes may be accomplished by creating another fish diagram with the possible cause now recorded in the "effect" box
- After condensing the possible causes, data is collected to verify the most probable causes; if the data does not substantiate the selected most probable causes, data is then collected on less obvious causes

Why-Why Pursuit

A divergent-convergent technique for pursuing the root cause for a problem where each divergent step in the why-why analysis is produced by asking *why?* The pictorial result is a horizontal tree depicting answers to the why questions. Procedure:

- Use brainstorming process as needed
- Record problem on board in a box with one line extending from the side of the box and a vertical line extending across the horizontal line
- Explore possible causes by asking *why?;* add a horizontal line from the vertical line for each answer to this why question and place each answer in a box at the end of a horizontal line
- Explore possible subcauses for each new box by asking *why?* and create a new column of boxes for these answers

- Continue asking *why?* for each new answer until a logical end is reached along each path; avoid superfluous or cyclical answers
- Convergent thinking techniques such as NGT or multivoting are used to focus on the most probable causes; convergent thinking techniques may also be used after each divergent step to narrow the list of alternatives before the next divergent step
- After condensing the possible causes, data is collected to verify the most probable causes; if the data does not substantiate the selected most probable causes, data is then collected on less obvious causes.

Example:

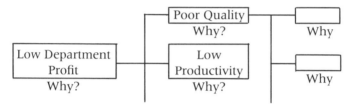

How-How Pursuit

A divergent-convergent technique for pursuing solutions where each divergent step in the how-how analysis is produced by asking *how?* This technique causes team members to creatively explore and consider numerous solution alternatives instead of jumping to the obvious solution. It is also useful in determining specific steps that should be taken to implement a solution. The pictorial result is a horizontal tree depicting answers to the how questions. Procedure:

- Use brainstorming process as needed
- Record the solution statement on board in a box with one line extending from the side of the box and a vertical line extending across the horizontal line
- Explore possible solutions by asking *how?;* add a horizontal line from the vertical line for each answer to this *how* question and place each answer in a box at the end of a horizontal line
- Explore possible subsolutions for each new box by asking *how?* and create a new column of boxes for these answers

- Continue asking *how?* for each new answer until a logical end is reached along each path; avoid superfluous or cyclical answers
- Convergent thinking techniques such as NGT or multivoting are used to focus on the most probable solutions; convergent thinking techniques may also be used after each divergent step to narrow the list of alternatives before the next divergent step
- Advantages and disadvantages, chance for success, and relative cost should be listed for each alternative to facilitate a more objective selection process

Force Field Analysis

This technique is extremely useful in solution finding and in solution testing and implementation. It is used to analyze the basic components of a problem, identify key elements of a problem about which something can be done, and develop a systematic strategy for problem solving which minimizes irrelevant efforts and inhibiting factors. It is used to illustrate the relative pros and cons of a solution, evaluate alternative solutions, and create a set of criteria for evaluation of action steps. Preface:

- Any problem situation may be thought of as an activity level which is different from that desired. The activity level results from pressures or influences acting upon the individual, group, or organization in question.
- These influences are called forces and there are two kinds of forces: driving and restraining. Driving forces promote the occurrence of the particular activity of concern. Restraining forces inhibit or oppose the occurrence of the activity. The two force fields push in opposite directions, and while one is stronger, a point of balance is usually achieved which gives the appearance of steady state (or habitual behavior).
- Changes in the forces of either field can cause a change in the Activity Level. Thus, apparently habitual ways of behaving can be changed and related problems solved by bringing about changes in the relative strengths of the driving and restraining force fields.

Procedure:

- The force field is depicted as follows:
  1. The activity level is depicted as a horizontal line

2. The intensity scale for restraining forces are indicated by equally spaced sequential numbers above the activity level line starting with the number 1

3. The intensity scale for driving forces are indicated by equally spaced sequential numbers below the activity level line starting with the number 1

• The activity level is defined and recorded next to the activity level line, for example, "Purchasing Department Personnel Attitudes"

• Divergent-convergent techniques are used to identify all possible driving and restraining forces

• Identified forces are discussed and the intensity of each force is judged and decided by group consensus (ranking techniques may be used)

• The forces are recorded on the force field diagram as vertical arrows pointing at the activity level line, with the lengths of the arrows depending on the relative strengths of the forces they represent; force names are recorded at the tail of each arrow

• Strategy is developed to maximize driving forces and minimize restraining forces

Example:

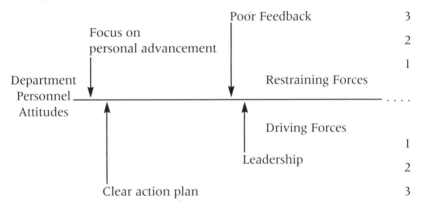

Process Analysis

This is a systems analysis technique which results in the development of a simplified process with a reduced set of tasks. In this technique, the current process is examined. Comprehensive interviews are used to define all tasks in a process, feedback loops, and functional interfaces; identify current problems and inefficiencies; and develop a flow diagram for the current process. This information is then used to develop a new flow diagram and define the new set of interrelated tasks.

Input-Process-Output Analysis

This is a systematic value-added approach to defining activities and tasks within an operation, generally at a department or functional level. This technique focuses on fulfilling customer needs, defining the activities which must be performed to satisfy the customers, defining inputs needed to perform the defined activities, and ensuring that the suppliers of the inputs are aware of and can fulfill these input needs. Procedure:

- Identify customer needs
- Identify current inputs and suppliers of the inputs
- Analyze current processes associated with fulfilling customer needs; this effort should include estimation of the value added to the final output of the process by each activity, and the time spent at each process step in performing work associated with adding value and with each area of quality cost (Internal Failure, External Failure, Appraisal, and Prevention)
- Redefine the process to maximize added value, minimize quality costs, and maximize efficiency of the process; this effort should include definition of needed inputs and suppliers of the inputs
- Work with input suppliers to ensure that input needs will be fulfilled

Annual Strategic Plans

The following are partial examples of annual strategic plans for the first year and several years later. Although some organizations tend to separate the strategic plan into two or more plans with one addressing quality improvement, the following examples reflect consolidated plans.

Communication Clinic 1981 Strategic Plan (first year of plan use)

Quality Policy: We will work together and with other members of the community to continually and enthusiastically pursue improvement and innovation in all areas of speech language pathology and interpersonal communication, and in all of CC's activities and services. We will endeavor to

* Improve the quality of life in the community by helping people to communicate more effectively and by our efforts to advance the art and science of communication,

- Improve the quality of our work life and support each other in improving the quality of our individual lives,
- Exceed the expectations of our external and internal customers.

Objectives and Implementation Plan

III. *Objective:* Enhance public awareness of the importance of effective communication and help the public learn to be more effective communicators.

    Plan: During 1981, we will produce and air forty "Communication Please" radio segments and pursue syndication of "Communication Please."

    Lead Responsibility: Director

    Schedule: Produce and air at least five new segments every six weeks. Complete an investigation into syndication and develop an action plan by September 1981.

IV. *Objective:* Enhance the level of satisfaction experienced by our contract clients.

    Plan: Identify contract client needs and expectations, obtain performance ratings on these characteristics for CC and our competitors, identify areas of CC weakness, and evaluate and improve CC processes which affect these weak characteristics.

    Lead Responsibility: Contract Service Liaison

    Schedule: Project report with recommended changes and implementation plans presented to staff by November 30, 1981.

VII. *Objective:* Enhance staff knowledge and professionalism.
. . .

IX. *Objective:* Expand and solidify current service programs.
. . .

X. *Objective:* Develop and implement systems for monitoring quality and productivity.

    Plan: Identify critical measures of quality and productivity. Complete development of, and implement, the CaseLoad monitoring system and the problem/concern action program. Complete implementation of the automated accounting system.

    Lead Responsibility: . . .

XII. *Objective:* (Re)define the organization and build administrative and management systems by December 1982 which will accommodate and facilitate growth locally and at satellite locations.

Plan: . . .

1981 Budget (not included in this Appendix)

Communication Clinic 1990 Strategic Plan (ninth annual plan)

Quality Policy: (same as 1981)

1989 Achievements

We continued our commitment to ongoing improvement in client satisfaction and in the effectiveness of our services at improving and stabilizing our client's communication abilities. We have achieved marked success in the area of client satisfaction with a 30% reduction of client related problems/concerns, 45% increase in client identified Exceptional Practices, and 5% increase in the very satisfied category on client surveys. Unfortunately, the improvements did not occur across the board as all satisfaction indicators experienced a statistically significant reduction at our Athens branch. (Actions to correct this situation are addressed under the heading "1990 Plan.")

Our efforts at enhancing the quality of our work life, professional and interpersonal skill, focus on positive mental attitude, and individual health resulted in noticeable success during the year. In particular,

- Participation at the health institute increased 50% bringing ongoing participation to 85%, and the health fitness tests for all participating associates indicated marked improvement.

. . .

1990 Quality Measures

The following seven quality measures have been selected for monitoring CC's overall quality health during 1990. These charts depict current status and objectives for 1990–1992.

```
Rating
          Customer Satisfaction
98           1990 Goal: 93%

96
                                  0000
94                           0000 0000
                        0000 0000 0000
92                XXXX  0000 0000 0000
                  XXXX  0000 0000 0000
90      XXXX      XXXX  0000 0000 0000
        XXXX      XXXX  0000 0000 0000

        ______________________________
        1988      1989  1990 1991 1992
```

What We Measure:
- Client identified concerns
- Concern recurrence
- Performance at meeting Client Needs
- Client identified Exceptional Practices

Why Is It Important:

How much our customers value our service, how effectively we meet their needs, our ability to quickly and effectively resolve customer concerns, and our ability to prevent recurrence are prime indicators of our performance.

. . .

1990 Plan

Our 1990 Plan is divided into five areas and fourteen characteristics of excellence.

Management/Leadership
- Culture
- Planning
- Communication

Service/Process
- Service Excellence
- Process Excellence
- Procedure Compliance
- Information

Human Resource
- Participation
- Skill Enhancement
- Health Enhancement

Customer Orientation
- Customer Satisfaction

Community Orientation
- Community Awareness
- Profession Advancement
- Ecological/Social Responsibility

. . .

The following are the specific goals and plans for obtaining the goals for each of the fourteen characteristics.

Service/Process
  Service Excellence
    Identify and establish a set of "Pulse Points" to measure effectiveness of our therapy procedures. Establish a tracking system and statistical tools for evaluation, and initiate tracking.
    (Tracking Number 90–212)
    Lead Responsibility: Clinical Supervisor
    Support: Assistant Director

Human Resource
  Participation
    All associates will establish personal and professional improvement goals and plans, and post them in their office area with progress charts. These goals will be incorporated into each associate's Performance Evaluation and Improvement Plan.
    (Tracking Number 90–311)
    Lead Responsibility: Director
    Support: Office Manager

. . .

1990 Budget (not included in this Appendix)

Policies and Programs Description Manual (Interim Manual)

The following is an example of selected portions of a policies and programs description manual. During your review of this manual and Appendixes IV and V, you may notice the similarity between the procedures and the equivalent portions of this manual. Before developing procedures, a comparison of these equivalent documentation approaches may provide you with valuable insights regarding the benefits and disadvantages of each approach, and help you understand the process of procedure development addressed in Appendix IV.

COMMUNICATION CLINIC: POLICIES AND PROGRAMS DESCRIPTION MANUAL
Number   Document Title

LONG-TERM FOCUS

0.1   Statements of Mission and Distinctive Competency

0.2   Objectives

STAFF GUIDELINES

1.01   Quality Focus

1.02   Image, Attitude, Ethics

1.03   Management and Organization

SERVICE PROGRAMS

2.01   Programs Management

2.02   Treatment

2.03   Client Administration and Satisfaction Management

SUPPORT PROGRAMS

3.01   Information Dissemination

3.02   Documentation Control

3.03   Policies and Programs Description Manual Control

3.04   Financial, Contractual, and Legal Management

3.05   Procurement Control

3.06   Equipment and Materials Control

3.07   Quality Concerns and Corrective Action

3.08   Evaluation, Innovation, and Improvement

3.09   Human Resource Management

3.10   Staff Benefits

3.11   Strategic Planning and Growth Management

Title: Documentation Control

P/PD Number: 3.02   Date: xx/xx/xx

Policy

Documentation shall be identified, developed, and maintained to provide guidance on important activities, objectively demonstrate effective operation of the administrative systems, attest to the quality of items and services, and aid in identifying, correcting, and preventing conditions adverse to quality.

Program Description

All documentation which is developed and/or maintained by CC falls into one or more of the following seven categories:

- Quality Assurance Records—A QA Record is an approved document which prescribes the manner in which an activity is to be performed, demonstrates that an activity was performed in accordance with applicable requirements, or provides primary evidence as to the quality of services or items. One copy of each QA Record is clearly marked "QA Record" and maintained for at least 4 years in a fire retardent facility with controlled access (the long-term storage facility).
- Supplemental QA Records—SQA Records are completed documents which provide supplementary evidence of quality or support the evidence contained in a QA Record, such as Daily Progress Notes. SQA Records are maintained in working files such as the central administrative files and individual employee files. SQA Records are maintained in the long-term storage facility when no longer needed for reference.
- Magnetic Media Documentation—An MMD is documentation which resides on magnetic media; e.g., correspondence and data bases which reside on computer storage devices and audio/video recordings of clients. Individual MMDs may be a QA or SQA Record, and are maintained accordingly; for example, training video tapes and tapes of client practice sessions.
- Policies and Programs Description—P/PDs are documents which provide guidance and control administrative, service and other activities, and are addressed in P/PD 3.03.

P/PD 3.02, Documentation Control (Page 2)

- Forms - These are documents which ensure the consistent collection of data and information, e.g., caseload survey. A form becomes a record once it is completed. Development and use of forms are controlled by and in a Forms Manual. The Forms Manual is maintained by the Records Clerk.
- Professional Materials - Many professional (service and administrative) materials are in the form of documentation, such as textbooks, test and equipment manuals, and journals. This documentation is addressed in P/PD 3.06.
- Working Documents - These are all other records and documents including copies of records and documents which are in the development stage such as draft reports. These documents are generally maintained by the individual employee until no longer needed.

CC's system for filing documentation is composed of the following five elements:

- Central Administrative Files—This facility is located in the administrative office. A chart of the file organization is maintained on the top of the cabinet. This facility is maintained by the Records Clerk.
- Individual Employee Files—These are files maintained by employees, such as desks
- Computer Files—The organization of the computer files is posted on the wall behind the computer and an index to the files is maintained next to the computer. Backup copies of the files are maintained for two years in the long-term storage facility. Unneeded files are removed from the hard disk at least twice a year. Accounting files are managed in accordance with software instructions.
- Bull-Pen Forms File—This facility contains copies of routinely used forms. It is located in the bull-pen and maintained by the Records Clerk.
- Long-Term Storage Facility—This is a fire retardent, limited access facility which contains the QA Records, SQA Records which are no longer needed on a day-to-day basis, and backup copies of the computer files. A chart of the file room's organization is posted. The Records Clerk is responsible for maintenance of this facility.

Policies and Programs Description Manual Control

P/PD Number: 3.03   Date: xx/xx/xx

Policy

The policies, organization, and administrative systems shall be documented in accessible form, and care shall be taken to enhance stability and ensure correctness of this documentation. Administrative, service, and other activities shall be prescribed in and controlled by documents to the extent that standardization or regulation of the activity is (1) necessitated by process complexity or repetition, or (2) important to the control, improvement, and assurance of service qualities. The resulting documents are not intended to replace training, experience, or abilities of the individual performing the activity, but to compliment their skills by providing the means for planning, organizing, and directing the work.

Program Description

CC's statements of mission, competency, and objectives; policies; organizational form; administrative systems; and service activities are documented in Policy and Program Description documents (P/PD) and controlled by this manual. Each P/PD has a format similar to this one, although either the policy or program description section may not be used. P/PDs are reviewed by members of the organization and approved by the Director. To ensure that current approved P/PDs are in use, manuals are issued to all members of the organization, subcontractors, and selected locations of use; each manual copy is uniquely numbered and assigned to one individual; and new and revised P/PDs are issued to the manual holders and are only valid when located within the manual.

Title: Procurement Control

P/PD Number: 3.05   Date: xx/xx/xx

Policy

Purchases of products and services shall be based on need as opposed to want. Product and service options and procurement sources shall be thoroughly investigated to ensure that options fulfill needs and that quality performance will be satisfactory. Suppliers shall be afforded respect and treated as members of the clinic team.

Program Description

The Office Manager is responsible for procurement decisions regarding equipment maintenance and supplies which are routinely used in the operation of the clinic. The Director is responsible for other procurement decisions. Procurement decisions are made in a timely manner based on:

- Need as evidenced by tangible benefits such as increased efficiency, improved customer service, increased morale and productivity, improved public image, etc.,
- Investigation of product/service needs,
- Investigation of suppliers, and their product/service options and quality performance history,
- Enough proposals to make an intelligent decision.

Where feasible, procurement investigations are performed by someone other than the person making the purchase decision, with the decision maker interviewing only the salespersons from the top one or two choices. (Minimizes the influence of the salesperson's personality.)

The "best possible price" is specifically requested, several times if necessary; for example, at the time the proposal is requested and during the closing negotiation.

The acceptability of the purchased products/services are assessed at the time of receipt. The needs and quality performance of the products/services and the supplier are reviewed prior to additional purchases.

Suppliers which have a consistent history of providing products/services which exceed clinic expectations are given preferential treatment and involved early in the procurement process.

Title: Quality Concerns and Corrective Action

P/PD Number:: 3.07   Date: xx/xx/xx

Policy

Every person working for CC is responsible for

- Providing a product or service to their customer (both external and the next person in the process) which conforms to applicable laws, regulations, and contractual and procedural requirements; and which exceeds the customer's expectations,
- Seeking ways to prevent quality problems,
- Remaining constantly alert toward the detection of quality problems and exceptional practices, including those associated with products and services received from outside CC,
- Being responsive to customers who are expressing a concern regarding the quality of CC's products and services, and
- Initiating and participating in the process of correcting problems and building upon exceptional practices.

The delivery and use of products and services reflecting a problem shall be controlled or precluded until the problem has been corrected. In addition, each problem which has or could have a major impact on CC's clients or operations shall be promptly reported to the Director, the problem and associated processes/systems shall be evaluated to determine the root cause of the condition, and action shall be taken to prevent recurrence.

Program Description

All staff members and contractors report the existence of problems and exceptional practices on a Comment and Concern Form which is also used to document the corrective action; and control identified problems to minimize any actual or perceived adverse impact on a customer. For confirmed problems, action is promptly taken to correct the problem, and where appropriate prevent recurrence. The Director ensures that all problems are evaluated and resolved in a timely manner, appropriate reccurence control is implemented, and exceptional practices are imbedded throughout the organization. The Director also reviews confirmed problems annually to identify recurrent problems, and initiates corrective action as appropriate.

Procedure System and Control (and Procedure Writers Guide)

This is an example of an extremely thorough procedure on "procedure system and control." It may be considered at the border of human ability to comprehend. However, in complex organizations, the benefits of such a procedure may outweight the potential difficulties of implementation.

This procedure and attached guide may best be used as a resource in studying and understanding the process for development and control of procedures.

Title: <u>PROCEDURE SYSTEM AND CONTROL</u>

Number: <u>P&P 3.03</u>   Revision: <u>0-DRAFT</u>

Issue Date: _____   Revision Date: _____

1 <u>PURPOSE AND APPLICABILITY</u>
To prescribe the responsibilities and controls for development, issuance, distribution, change, and maintenance of procedures. This procedure applies to all procedures issued by the Communication Clinic.

2 <u>REFERENCES AND RESPONSIBILITIES</u>
2.1 P&P 0.2, OBJECTIVES, Numbers 1 and 4
2.2 The following positions/functions are assigned responsibilities within this procedure. Responsibilities assigned to titled positions/functions may be delegated unless stated otherwise; however, the person filling the titled position/function retains ultimate responsibility.
   • Director
   • Office Manager
   • Records Clerk
   • all Associates

3 <u>ABBREVIATIONS AND DEFINITIONS</u>
3.1 CC -Communication Clinic
3.2 OP -Office Procedures
3.3 OPM -Office Procedures Manual
3.4 P&P -Policy and Program Procedures
3.5 P&PM -Policy and Program Procedures Manual
3.6 T&D -Treatment and Documentation Procedures
3.7 T&DM -Treatment and Documentation Procedures Manual
3.8 May/Should -The words "may" and "should" are used to indicate generally appropriate actions, or guidance, in the given situation. This guidance should be followed unless there are compelling reasons to take other actions.
3.9 Will/Shall—The words "will" and "shall" are used to identify required actions. Other actions are not acceptable. If the required action is not appropriate, change the procedure (see paragraph 4.4.6).

4 <u>POLICY/PROCEDURE</u>
4.1 <u>Procedure System Policies</u>
   4.1.1 The policies, organization, and administrative systems shall be documented in accessible form, and care shall be

taken to enhance stability and ensure correctness of this documentation.

4.1.2 Administrative, service, and other activities shall be prescribed in and controlled by documented procedures. These activities shall be prescribed to the extent that standardization or regulation of the activity is (1) necessitated by process complexity or repetition, or (2) important to the control, improvement, and assurance of service qualities. The resulting documents are not intended to replace training, experience, or abilities of the individual performing the activity, but to compliment their skills by providing the means for planning, organizing, and controlling the work.

4.2 Procedure System Overview

4.2.1 **Procedure System Description:** CC's statements of mission, competency, and objectives; policies; organizational form; administrative systems; and service activities are documented in and controlled by the manuals listed below. Each manual addresses a specific scope of CC's activities, and the procedures in each manual conform to a uniform format. To ensure that current approved procedures are in use, manuals are distributed to those who are required to use them or to the location of use; each manual copy is uniquely numbered and assigned to one individual; and procedures are issued to the manual holders and are only valid when located within the appropriate manual. CC's manuals and scope of applicability are as follows:

a. Policy and Program Procedures Manual—The P&PM documents the statements of mission, competency, and objectives; defines the organization; and establishes the policies, management and service programs, administrative systems, and staff guidelines. The Director is responsible for the form and content of this manual.

b. Treatment and Documentation Procedures Manual—The T&DM controls the technical aspects of CC's services and the documentation of those services. It prescribes the requirements, guidance, and where appropriate, detailed instructions for the service departments. The Director is responsible for the form and content of this manual.

c. Office Procedures Manual—The OPM contains the detailed instructions for performance and documentation of office functions (such as backing up the computer files and closing the general ledger). The Office Manager is responsible for the form and content of this manual.

4.2.2 **Index to Procedure Section:** The following is an index to Section 4 of this procedure.

4.1 Procedure System Policies

4.2 Procedure System Overview

    4.2.1 Procedure System Description

    4.2.2 Index to Procedure Section

4.3 Procedure Manuals

    4.3.1 Distribution and Control

    4.3.2 Manual Format

    4.3.3 Procedure Organization and Numbering

4.4 Procedures

    4.4.1 Preparation

    4.4.2 Format

    4.4.3 Review, Concurrence, and Approval

    4.4.4 Distribution

    4.4.5 Revision

    4.4.6 Limited Duration Revision Memorandum

    4.4.7 Periodic Review

4.3 Procedure Manuals

4.3.1 **Distribution and Control:** Each copy of the P&P, T&D, and OP manuals shall be uniquely numbered and assigned to a specific individual as indicated below. The appropriate copy(s) shall be transmitted to each assigned person, and the distribution lists shall be maintained by the Records Clerk. The individual manual holders are responsible for ensuring the security and integrity of their manuals. Each time a procedure is issued but at least once each year, the Records Clerk shall issue a copy of the current manual index to each manual holders along with a return-acknowledgment transmittal. Each manual holder shall attest in writing on the transmittal that their manual is complete and all procedures are of the current revision, and shall return the transmittal within seven days of receipt.

a. Policy and Program Procedures Manual—P&PMs shall be issued to each person with management responsibilities, each person assigned specific responsibilities in the manual, and others as designated by the Director. As a minimum, at least one additional copy shall be located in the Bull Pen and in the Administrative Office, and manual holders shall be designated for these manuals.

b. Treatment and Documentation Procedures Manual—T&DMs shall be issued to each person providing service and others as designated by the Director. As a minimum, at least one additional copy shall be located in the Bull Pen and in the Administrative Office, and manual holders shall be designated for these manuals.

c. Office Procedures Manual—OPMs shall be issued to each person performing office functions and to others as designated by the Office Manager. As a minimum, at least one additional copy shall be located in the Bull Pen and in the Administrative Office, and manual holders shall be designated for these manuals.

4.3.2 **Manual Format:** The manual title shall be clearly visible on the binder of each manual. All manuals of the same type (e.g., all P&PMs) shall have the same color binder. The first page or cover of each manual shall contain the manual title, unique copy number, and manual holder's name. The next page shall be an index of the procedures contained in the manual. The index shall identify the procedures by number, revision, and date of latest issuance. The index shall also identify the date of index revision. The procedures shall follow the index in numerical order. The individuals designated in 4.2.1a–c are responsible for format of the respective manuals.

4.3.3 **Procedure Organization and Numbering:** Each manual shall be divided into sections. Each section of the manual shall be assigned a whole number (e.g., 1, 2, . . .) and each procedure within a section shall be assigned a decimal number (e.g., 1.01, 1.02, 1.03, . . .). The individuals designated in 4.2.1.a through c are responsible for manual organization and procedure numbering.

4.4 Procedures

4.4.1 **Preparation:** Procedures may be drafted by anyone with responsibility and experience in the procedural area. The individuals designated in 4.2.1.a through c are responsible for designating preparers for procedures within the respective manuals. Additional guidance is provided in attachment 6.1.

4.4.2 **Format:** The statements of mission, competency, and objectives shall be typed on letterhead and shall clearly identify the P&P number, title, revision, issue date, and approval signatures. All other procedures shall be typed on the appropriate form and contain the sections as identified on attachments 6.2–6.4. Additional guidance is provided in attachment 6.1.

4.4.3 **Review, Concurrence, and Approval:** Every procedure shall be reviewed for adequacy, clarity, correctness, and completeness. Review comments shall be given due consideration and the individuals making comments which are rejected shall be informed of the reason for rejections. The procedure shall then be redrafted and rereviewed, as necessary. Once the comments are resolved and the procedure is acceptable, the originator, independent reviewer, concurrer(s), and approver(s) shall sign the document.

a. Each procedure should be reviewed by individuals performing, or responsible for, the activities addressed in the procedure.

b. Each procedure shall be reviewed by an independent reviewer. (An independent reviewer is someone with experience in the procedural area other than the preparer and approver.) This reviewer has the additional responsibility for assuring that all contract, regulatory, industry, company and other requirements applicable to the procedural area are appropriately addressed in the procedure; that the requirements, guidance, and processes addressed in the procedure are consistent with those imposed by related procedures; and that the appropriate Associates have reviewed the procedure.

c. Concurrence is required from affected management positions whenever the procedure impacts a segment of CC's organization not under the scope of the procedure (e.g., when a T&D imposes requirements on the office staff). Concurrence signatures attest to their commitment to implement the interfacing aspects of the procedure.

d. The individuals designated in 4.2.1.a through c are responsible for identifying reviewers and concurrers and for approving the procedures. Approvers have the additional responsibility for assuring compliance with this procedure; ensuring that all procedures within their area of responsibility are uniform in format and style; and ensuring that overlaps and gaps between procedural coverage are minimized. Approval signatures attest to their commitment to ensure implementation of the procedure.

e. Statements of Mission, Competency, and Objectives should be reviewed and concurred with by all members of CC's staff; however, only the approval signature shall be placed on these statements.

4.4.4 **Distribution:** Procedures shall be transmitted by the Records Clerk along with a revised manual index to all holders of the applicable manual. The original copy of the procedure and the index shall be placed in QA Record storage by the Records Clerk. The transmittal notice shall be signed within 7 work days by the manual holders and returned to the Records Clerk. This signature shall document that the procedure(s) and index have been placed in the appropriate manual, superseded procedure(s) and index have been removed from the manual, all procedures and only those procedures identified on the index are in the manual, and all procedures in the manual are of the correct revision level. Manual holders are responsible for correcting identified deficiencies.

4.4.5 **Revision:** Procedures shall be revised as necessary to ensure adequacy, clarity, completeness, and correctness. The individuals designated in 4.2.1.a–c are responsible for procedural revisions. Potential inadequacies in procedures or manuals shall be reported to and evaluated by these individuals. Revisions shall be drafted, reviewed, concurred and approved in accordance with the requirements of paragraphs 4.4.1–4.4.4. Revised lines of text shall be marked by a line or dagger in the right hand margin.

4.4.6 **Limited Duration Revision Memorandum:** On infrequent occasions, it may be prudent to not fulfill specific procedural requirements or impose additional requirements for a limited time or for a specific project. This is accomplished by issuance of a Limited Duration Revision Memorandum (LDRM)

a. The individuals designated in 4.2.1.a–c are responsible for issuing and cancelling LDRMs.

b. LDRMs shall be developed, reviewed, approved, and distributed in accordance with the requirements of paragraphs 4.4.1, 4.4.3, and 4.4.4. LDRMs shall be placed in front of the procedure which they modify.

c. LDRMs shall clearly indicate the initiation and termination dates/events, and the requirements which are being modified.

d. LDRMs shall be uniquely numbered as follows: procedure identifier, number, and revision—LDRM identifier and sequential number (e.g., P&P 3.03 R1 - LDRM 5).

e. LDRMs shall have the same title as the procedure which the LDRM is modifying.

f. LDRMs shall be canceled and removed from the manual by issuance of a revised procedure, but no later than the termination date. This shall be accomplished by issuance of a revised manual index in accordance with paragraph 4.4.4.

4.4.7 **Periodic Review:** All procedures and manuals shall be reviewed at least every 2 years to ensure adequacy, clarity, completeness, and correctness. The individuals designated in 4.3.1.a–c are responsible for ensuring that periodic reviews are performed and weaknesses resolved.

5   RECORDS

5.1 QA Records

The following are QA Records which are required by this procedure and maintained in accordance with P&P 3.04.

a. All approved procedures and revisions (P&Ps, T&Ds, OPs)

b. Indices and revisions for all manuals (P&PM, T&DM, OPM)

c. Lists of manual holders for all manuals (P&PM, T&DM, OPM)

d. Limited Duration Revision Memorandums

5.2 Other Records

5.2.1 **Supplemental QA Records:** Receipt-Acknowledgement Transmittals are maintained in the central administrative files by the Records Clerk for at least the most recent two transmittals for each manual. These records are disposed of upon removal from these files.

5.2.2 **Archive Files:** Previous revisions of procedures, manual indices, distribution lists, and limited duration revision memorandums are to be removed from the hard disk and placed on archive floppy disks at the time the new revisions are issued. These archive floppy disks shall be maintained in accordance with P&P 3.04, Documentation Control.

6 <u>ATTACHMENTS</u>
6.1 Guidance on Drafting Procedures
6.2 P&P Form and Format
6.3 T&D Form and Format
6.4 OP Form and Format

7 <u>PREPARATION, REVIEW, CONCURRENCE, APPROVAL SIGNA-
TURES</u>

Attachment 6.1
P&P 3.03, Revision 0

<u>GUIDANCE ON DRAFTING PROCEDURES</u>

1 <u>PURPOSE AND APPLICABILITY</u>
To provide guidance for the preparation of procedures. This guidance is applicable to all procedures issued within the Communication Clinic.

2 <u>ABBREVIATIONS AND DEFINITIONS</u>
2.1 P&P—Policy and Program Procedures

3 <u>STYLE AND DEVELOPMENT CONSIDERATIONS</u>
3.1 <u>General</u>
Procedures are the cornerstone of a quality system. It is important that they be designed to serve people who must implement them. It is rare that a procedure is read through more than once. After the first reading, it becomes a reference. Therefore it is important to design the procedure that way: information should be easy to find and all the information on one subject should be in one place. This requires planning.
3.2 <u>Planning</u>

Before you write, outline. List all the points you want to cover, and then rearrange them in a logical order. Analyze the outline carefully to be sure that each step leads naturally to the next, that there are no redundant steps, that there are no gaps in the chain of events, and that the process is efficient, effective, and will lead to meaningful results. Once you have outlined the process, it is time to fit that material into the format of the written procedure; refer to sections 4 and 5 of this attachment for guidance on procedure and attachment format.

3.3 <u>Style</u>

3.3.1 **Triggers:** Each activity should lead into the next; ideally, the end of one activity should serve as the "trigger" for the next one. A trigger is the event that tells a person that some activity must be started. Triggers should be clearly identified in the procedure.

3.3.2 **Loop Closure:** Another characteristic of a good procedure is loop closure. This is the feedback that tells an individual that an activity was completed or a step accomplished. A "receipt-acknowledgment" transmittal is an example of loop closure.

3.3.3 **Clarity and Precision:** Just as important as the content of the procedure is the precision and clarity of the writing.

- Precise wording is the first requirement. Be sure that your words say exactly what you mean and that they cannot be misconstrued. Use words with specific meanings rather than words that could be interpreted differently by different people. Avoid requiring subjective interpretation. For example, if you say an individual may be recertified if performance has been satisfactory, you are asking many people to decide what is "satisfactory."
- When you are working on a procedure, you cannot always anticipate the circumstances under which it will be referenced. The only alternative is to make every part crystal clear.

3.3.4 **Balance:** Procedures can cause more damage than good if the proper balance between requirement and guidance is not maintained. Overly restrictive procedures inhibit individual creativity and ability to effectively deal with unusual situations. Excessive reliance on guidance creates opportunities for problems. Good procedures have a balance which is appropriate to the specific application and situation, and lean towards restrictive detail only when success of the activity depends on a specific sequence of precise actions.

3.3.5 **Initials and Acronyms:** The use of initials and acronyms to avoid the repetition of long titles is acceptable as long as it is not overdone. Consider these guidelines:
- Don't use initials for one-word titles,
- Use initials for two-word titles only if the words are long (e.g., 14 letters total),
- Don't use initials if they could have multiple meanings,
- Don't use initials for a title if it appears only once or twice in the procedure.

3.4 Test-Use and Debug

3.4.1 **General:** Due to the complexity of some processes and activities, the initial version of some procedures may contain several inefficient requirements or errors. Procedure test-use and debugging is a way to ensure that a procedure is adequate before formal issuance, and avoid requiring people to comply with a ridiculous procedure and wasting precious effort and resources.

3.4.2 **Process:** Before formal issuance of a procedure but after reviews, concurrences, and approvals have been informally obtained, the procedure may be test-used and debugged as following:
- A clean draft of the procedure is printed.
- The person(s) responsible for approving the procedure documents the scope of the test-use on the draft, and initials the draft to document preliminary approval.
- The draft is used to perform the activities.
- Lessons learned during the test-use are studied to improve the procedure and the draft is revised to eliminate identified weaknesses.
- The revised draft is again reviewed by the people who reviewed the previous draft and issued through the normal process.

3.4.3 **Example—Case Load Survey:** Prior to issuing a revised case load survey procedure, the decision is made for a couple of pathologists to use the new procedure for 2 weeks to ensure that the process is fully developed, guidance is adequate, and the results are meaningful.

4 GUIDANCE ON PROCEDURE FORMAT

4.1 Procedure Identification

    4.1.1 **Title Section:** Every procedure starts with the procedure title, number, revision, date of initial issue, and date of revision. This is depicted on P&P 3.03 Attachment 6.2.

    4.1.2 **Header & Footer:** Every page of every procedure has a header which contains the company name and manual title; and a footer which identifies the procedure number and revision, current page number, and total number of pages in the procedure. The footer also contains the word "DRAFT" and the date during procedure preparation, review, and comment resolution. This is depicted on P&P 3.03 Attachment 6.2.

4.2 Section 1, "PURPOSE AND APPLICABILITY"

This section describes the reasons for writing the procedure and the functions or extent of the organization to which the procedure applies. This section should consist of a single paragraph.

4.3 Section 2, "REFERENCES AND RESPONSIBILITIES"

This section lists the documents which prescribe the requirements that are implemented in the procedure, such as a Statement of Objective, other procedures, regulations, and industry standards. This section also identifies the positions/functions which are assigned responsibilities within the procedure.

4.4 Section 3, "ABBREVIATIONS AND DEFINITIONS"

This section contains the definitions for acronyms and words with special meaning which are used in the procedure. Words already defined in other procedures should not be redefined, if possible. Having the same word defined in two different places invites conflict. This section contains the word "None" when there are no abbreviations and definitions.

4.5 Section 4, "POLICY/PROCEDURE" and "PROCEDURE"

    4.5.1 **Policy:** The first topic in this section documents the policies which direct the activities and decisions within the procedural area. There is a policy subsection when the referenced Objectives and P&Ps do not provide adequate guidance to ensure long-term consistency of purpose in the procedural area.

    4.5.2 **Overview:** The next topic in this section provides an overview of the procedural area. This subsection describes the major aspects of a program or provides an index to a lengthy and diverse procedure section. It does not restate policies or requirements, and may be omitted for simple and short procedures which do not control major program or functional areas.

    4.5.3 **Procedure:** The remainder of this section contains the actual procedure being documented and controlled. It prescribes the process, responsibilities, and detailed instructions, and provides the guidance.

4.6 Section 5, "RECORDS"

This section identifies the records which are required to be maintained by the procedure. It also specifies where and for how long they are to be maintained when the general requirements of P&P 3.04 are not adequate. It contains two subsections, QA Records and Other Records. These subsections contain the word "None" when no records are required.

4.7 Section 6, "ATTACHMENTS"

This section lists attachments in the order in which they are mentioned in the text. Attachments may be illustrations, copies of forms which are required by the procedure, instructions for completion of forms, or process flow charts, or may provide additional guidance which would unnecessarily complicate the procedure. This section contains the word "None" when there are no attachments.

4.8 Section 7, "PREPARATION, REVIEW, CONCURRENCE, AND APPROVAL SIGNATURES"

This section contains the signatures of the individuals who prepared the procedure, performed the independent review, are concurring with the procedure, and are approving the procedure.

4.9 Numbering, Indention, and Highlighting Conventions

    4.9.1 **Title Section:** The information provided in the procedure title section is presented in bold, all capital, underlined print.

    4.9.2 **Subsequent Sections:** Each subsequent section is identified by a header which starts with a single integer (e.g., 4) and contains the section title. The section title is presented in bold, all capital, underlined print.

    4.9.3 **Subsections:** Subsections in the POLICY/PROCEDURE and RECORD sections are identified by an indented header which starts with a number containing one period (e.g., 4.3)) and is followed by the subsection title. The subsection title is presented in first letter capital, underlined print. The other sections of the procedure contain lists, not subsections, see paragraph 4.9.7 below.

4.9.4 **Paragraphs -** POLICY/PROCEDURE and RECORD Sections: Paragraphs are started with a number containing two periods (e.g., 4.3.2), and a paragraph topic and colon are presented in bold first letter capital print (this is an example of a paragraph). The number is placed under the subsection title and the paragraph is indented with the first letter in the topic. The following are exceptions to this format.

    a. Paragraph is not numbered in subsections with only one paragraph are not numbered and the subsection header identifies the topic.

    b. Policy paragraphs do not contain a topic identifier.

    c. An overview subsection which only provides an index to the procedure section may be treated as a single paragraph followed by the index, where the index identifies the subsection headers, paragraph topics, and subsection and paragraph numbers.

4.9.5 **Sub-Paragraphs:** When further subdivision is needed in a paragraph, sub-paragraphs are used. Sub-paragraphs begin with a small letter followed by a period (e.g., a.) and are indented with the first letter of the first word in the paragraph (paragraph 4.9.4 above contains sub-paragraphs).

4.9.6 **Lists Within a Paragraph or Sub-Paragraph:** Each item in the list is started with a period and indented with the first letter of the first word of the item. There should be no further subdivision of information.

4.9.7 **Other Lists:** The REFERENCES AND RESPONSIBILITIES, ABBREVIATIONS AND DEFINITIONS, and ATTACHMENTS sections contain lists of information. Each item in the list starts with a number containing one period (e.g., 2.1) and the indention starts with the first letter of the item description.

5 GUIDANCE ON PROCEDURE ATTACHMENTS

   5.1 General

Attachments may be illustrations, copies of forms which are required by the procedure, or process flow charts, or may provide instructions for completion of forms or additional guidance which would unnecessarily complicate the procedure (this attachment is an example of a guidance type attachment). Attachments do not prescribe requirements or responsibilities but may define the process or provide the detail guidance by which requirements are fulfilled, such as instructions for completion of forms.

5.2 Attachment Identification—Forms and Illustrations
The upper right of each page contains the following information in block form as depicted on P&P 3.03 Attachment 6.2:
- Attachment number
- Procedure number and revision
- Attachment title in all capitals
- Current and total number of pages in the attachment.

5.3 Attachment Identification—Guidance and Flow Charts
The attachment number, procedure number and revision, and attachment title are centered on the top of the first page of the attachment with the title printed in bold, all capitals, and underlined. Each page of the attachment contains a footer which identifies the procedure number and revision, attachment number, and current and total number of pages in the attachment. This attachment is an example of a guidance attachmment.

5.4 Attachment Format—Forms, Examples, and Flow Charts
There are no format guidelines for these documents.

5.5 Attachment Format—Guidance Type
5.5.1 **Contents:** These attachments should start with a PURPOSE AND APPLICABILITY section and, where appropriate, REFERENCES and ABBREVIATIONS AND DEFINITIONS sections (see Sections 4.2, 4.3, and 4.4 of this attachment). The actual guidance may be provided in one or more appropriately titled sections. A guidance type attachment should not contain a RECORDS section because it should not prescribe requirements, or a signature section because it is endorsed and approved by the procedure of which it is a part. A guidance type attachment should not contain other attachments because of the inherent confusion, but may contain appendices if absolutely necessary. When used, appendices should be listed in a section titled APPENDICES and be listed in the order referenced in the attachment.

5.5.2 **Numbering, Indention, and Highlighting Conventions:** See paragraphs 4.9.2–4.9.7 of this attachment for the applicable conventions. Note that the POLICY/PROCEDURE section is replaced by appropriately titled sections containing the guidance.

Communication Clinic	POLICY AND PROGRAM PROCEDURES MANUAL

Attachment 6.2
P&P 3.03, Revision 0-D
P&P FORM AND FORMAT
Page 1 of 2

Title: [title]
Number: P&P [#.##]   Revision: [#]
Issue Date:_____ Revision Date:_____

1 PURPOSE AND APPLICABILITY
2 REFERENCES AND RESPONSIBILITIES
   2.1 P&P 0.3, OBJECTIVES, Numbers [#, #, and #]
   2.2 P&P [#.#] [title]
   2.3 The following positions/functions are assigned responsibilities within this procedure. Responsibilities assigned to titled positions/ functions may be delegated unless stated otherwise; however, the person filling the titled position/function retains ultimate responsibility.
      • [position]
3 ABBREVIATIONS AND DEFINITIONS
   3.1 OP—Office Procedures
4 POLICY/PROCEDURE
   4.1 (procedure area) Policies
   4.2 (procedure area) Overview
   4.3 [section title]
      4.3.1 [topic]: [paragraph]
         a. [subparagraph]
            • [listing]

Communication Clinic	POLICY AND PROGRAM PROCEDURES MANUAL

Attachment 6.2
P&P 3.03, Revision 0-D
P&P FORM AND FORMAT
Page 2 of 2

5 <u>QA RECORDS</u>
   5.1 <u>QA Records</u>
   5.2 <u>Other Records</u>
6 <u>ATTACHMENTS</u>
   <u>6.1</u>
   6.2
7 <u>PREPARATION, REVIEW, CONCURRENCE, APPROVAL SIGNA-
TURES</u>

Quality Concerns and Corrective Action (Procedure)

Title: QUALITY CONCERNS AND CORRECTIVE ACTION

Number: P&P 3.08   Revision: 0

Issue Date: _____   Revision Date: _____

1 PURPOSE AND APPLICABILITY

To prescribe the responsibilities and controls for handling and correcting problems, preventing recurrence of those problems which have a major adverse impact on customer satisfaction or business/professional operations, and identifying and building upon practices which have a major positive impact on customer satisfaction and business/professional operations. This procedure applies to all clinic staff and contract Associates.

2 REFERENCES AND RESPONSIBILITIES

2.1 P&P 0.2, OBJECTIVES, Numbers 1 and 5

2.2 P&P 1.1, QUALITY FOCUS

2.3 The following positions are assigned responsibilities within this procedure. Responsibilities assigned to titled positions may be delegated unless stated otherwise; however, the person filling the position retains the ultimate responsibility.
- Director
- All staff members
- All contract personnel.

3   ABBREVIATIONS AND DEFINITIONS

3.1 CC - Communication Clinic

3.2 C&C - Comment and Concern

3.3 Exceptional Practice—Activities, services, and products which fulfill the customer's needs and cause the customer's expectations to be exceeded.

3.4 Quality Problem - A situation which presents uncertainty, perplexity or difficulty regarding quality, including a condition which appears to have the potential for causing the quality of an item, service, or activity to be unacceptable or questionable.

3.5 Quality—The totality of features and characteristics of a product or service that bears on its ability to satisfy stated or implied needs. It encompasses safety, performance, dependability, effectivity, timeliness, cost, and productivity of the product or service and the activities associated with production of the product or service.

3.6 SQP - A Significant Quality Problem is a situation which violates or appears to violate a law, regulation, or contractual requirement, holds the potential for causing a strongly negative impact on a client's perception of CC, or holds the potential for causing a strongly negative impact on CC's business/professional operations.

4   POLICY/PROCEDURE

4.1 Quality Concern and Corrective Action Policies

4.1.1 Every person working for CC is responsible for
- Providing a product or service to their customer (both external and the next person in the process) which conforms to applicable laws, regulations, contractual and procedural requirements; and which exceeds the customer's expectations,
- Seeking ways to prevent quality problems,
- Remaining constantly alert toward the detection of quality problems and exceptional practices, including those associated with products and services received from outside CC,

- Being responsive to customers who are expressing a concern regarding the quality of CC's products and services, and
- Initiating and participating in the process of correcting quality problems and building upon exceptional practices.

4.1.2 The delivery and use of products and services reflecting a quality problem shall be controlled or precluded until the problem has been corrected. In addition, each problem which has or could have a major impact on CC's clients or operations and the associated processes/systems shall be evaluated to determine the root cause of the condition and action taken to prevent recurrence.

4.2 Quality Concern and Corrective Action System Overview

4.2.1 **System Description:** All CC staff members and contractors are required to remain alert to and identify the existence of quality problems and exceptional practices, initiate the corrective action process, and control identified problems to minimize any actual or perceived adverse impact on a customer. For confirmed quality problems, action is promptly taken to correct the problem, and action is taken to prevent recurrence of significant and recurring problems. The Director ensures that all quality problems are evaluated and resolved in a timely manner, appropriate corrective action is taken, and exceptional practices are imbedded throughout the organization. A Comment and Concern Form is used to initiate and document this process.

4.2.2 **Index to Procedure Section:** The following is an index to Section 4 of this procedure.

4.3 General Requirements

    4.3.1 **Director Notification:** The Director shall be notified of all SQPs within 24 hours of detection. This notification should be made by the person detecting the problem.

    4.3.2 **Release of Services/Products:** Product and services reflecting a quality problem shall not be provided to a client or contractor unless authorized by the Director. This authority shall not be delegated when the problem is a SQP.

4.4 Detection and Response

    4.4.1 **General:** The initial indication of a quality problem or exceptional practice may come from any source, including review of records, recognition of a customer's misunderstanding, and customer comment/complaint. In all cases, CC staff members and contract personnel must respond to the indication with an honest concern for improving quality and with a determination for achieving customer satisfaction.

    4.4.2 **Specific Actions:** The person detecting the quality problem or exceptional practice shall take the following actions.

        a. Take such action as is deemed appropriate to ensure that the product(s) or service(s) reflecting quality problem is not released to an outside customer until the problem is resolved or the Director authorizes release.

        b. Obtain a clear understanding of the problem/practice, and to the extent feasible, the apparent causes for the problem.

        c. Immediately notify the Director if the problem is a SQP.

        d. When practical and within the authority of the person detecting the problem, correct the problem.

        e. Promptly initiate a C&C Form (Attachment 6.1) and place it in the C&C In-Box located in the administrative office. As an alternative for Associates and contract personnel who are not at the administrative office when the C&C requires prompt action, the following information may be telephoned into the administrative office and the form initiated by the person receiving the information:

- Details of the problem/practice, including the name of the person who detected the problem/practice and the name(s) of clients affected by the problem,
- Actions taken to correct the problem and control release of products/services reflecting the problem,
- Timing restraints for problem correction and release of services/products.

4.5 C&C Evaluation and Resolution

4.5.1 **General:** The Director should check the C&C In-Box and initiate resolution each day but at least once each week; this action should be delegated to the Assistant Director or Office Manager when the Director is unavailable. The Director is responsible for directing the evaluation and resolution process and closure of the C&C. The Director is also responsible for ensuring that the C&C is evaluated and resolved in a timely manner consistent with its impact on client satisfaction and CC operation.

4.5.2 **C&C Identification:** Each C&C shall be assigned a unique tracking number based on the date of initiation. For example, Y.MM.DD. ## where Y is the last digit of the year, MM is the month, DD is the day, and ## is a sequential number which starts as 01 each day.

4.5.3 **Resolution Process:** The following is a general discussion of the resolution process.

a. The Director assesses the impact of the C&C on client satisfaction and CC operation, identifies the person who is to be responsible for evaluation and resolution (hereafter referred to as the Responsible Person), provides guidance on the resolution process, defines the minimum scope of the evaluation, and defines the authority of the Responsible Person.

b. There are two distinct stages to the resolution process for C&Cs which involve quality problems

- Remedial Action;—fixing the product/service so that it may be provided to the client.
- Recurrence Control—determining the underlying causes of the problem including evaluation of associated processes/systems, taking action to prevent recur-

rence, and assessing the effectivity of the corrective measures. Recurrence control is required when the problem is an SQP or otherwise considered to have a major impact on client satisfaction or CC operation.

c. The resolution process for C&Cs which involve exceptional practices focuses on identifying circumstances when the exceptional practice is not used and instituting changes to ensure consistent application of the exceptional practice.

d. The evaluation should include interviews with people involved in the problem/practice and others generally associated with production and delivery of the service/product. The evaluation of SQPs and exceptional practices should also include the use of statistical methods for assessing the nature of the processes and determining appropriate changes.

e. People substantially impacted by the corrective measures should be consulted prior to implementation of the measures.

f. The Director should be very clear regarding the Responsible Person's authority to implement corrective measures without explicit approval of the measures by the Director.

g. The C&C form must be completed as the process progresses and returned to the Director for closure.

h. The person who detected the problem/practice must be notified of the resolution following closure of the C&C.

4.6 C&C Tracking

The Director shall maintain a log of C&Cs and ensure that C&Cs are resolved in a timely manner. The log is to be used to track the corrective action process and identify quality trends. The log should:

• Identify the unique C&C number,
• Indicate whether the C&C addresses a quality problem, SQP, or exceptional practice,
• Describe (briefly) the problem/practice, including affected client(s) and personnel,
• Indicate the current corrective action status, including the person assigned responsibility for investigating/resolving the C&C.

4.7 C&C Trend Evaluation and Quality Monitoring Report

During the first quarter of each year, the Director shall review all C&Cs, identify quality trends, including any unacceptable repetition of quality problems. The director shall then issue a Quality Monitoring Report which briefly addresses the following:

- Identified major quality problems (including problems identified as being repetitive), actions taken (or planned) to correct the problem and preclude recurrence, and the status of these actions,
- Identified exceptional practices and actions taken (or planned) to institutionalize the practices within CC, and the status of these actions,
- Identification of negative quality trends, actions taken to reverse the trend, and the status of these actions,
- Identification of positive quality trends, actions taken to ensure continuation of the trend, and the status of these actions,
- Evaluation of the effectiveness of the Quality Concern and Corrective Action process.

5 RECORDS

5.1 QA Records

The Quality Monitoring Report and completed C&C Forms are QA Records which are controlled by this procedure and maintained in accordance with Policy and Program procedure 3.04.

5.2 Other Records

Supplemental QA Records: The C&C Log is a Supplemental QA Record which is controlled by this procedure. Each page of this log shall be maintained by the Director for at least three years following closure of the last C&C on the page. These records may be disposed thereafter.

6 ATTACHMENTS

6.1 Comments and Concerns Form

7 PREPARATION, REVIEW, CONCURRENCE, APPROVAL SIGNATURES

COMMENTS AND CONCERNS

Attachment 6.1
P&P 3.08, Revision 0
COMMENTS AND CONCERN
Page 1 of 2

Initiator: _____ Date: _____

Response Time: _____ Immediate _____ ASAP

Comment/Concern (inc, impact & customer name, where applicable):

_____

_____

_____

_____

_____

_____

_____

_____

_____

_____

INITIATOR - DO NOT WRITE BELOW THIS LINE

Tracking Number: _____

Concern Evaluation & Correction - Assigned to: _____
Mgr Initials/Date: _____
(see back for directions on evaluation and correction)

Closure - Mgr. Initials/Date: _____

Initiator Acknowledgement (Initials/Date of initiator):

_____

Tracking Log Updated (Initials/Date of the person updating the log):

_____

Attachment 6.1
P&P 3.08, Revision 0
COMMENTS AND CONCERNS
Page 2 of 2

Evaluation & Correction Directions/Results:

NOTE: Attach additional documentation of the results of the evaluations, corrective actions, and required approvals.

Comment/Concern Trend Categories

A. Initiator: Customer ____ or Staff ____

B. Choose one: 1. Comment (Positive Feedback) ____
                2. Product/Service/Material/Equipment (PSME) not meeting expectations ____
                3. PSME needed but not being provided ____
                4. PSME being provided but not needed ____

C. Changes made to: ____ Process ____ Material/Equipment ____Product/Service Requirements/Specifications ____Other

Bibliography

The ideas and information presented in this book were compiled from years of experience, study, and research. The following list reflects the published documents which were specifically referenced during the development of this work.

Alexander, C. P. "Quality and the Inner Enterprise," published in the June 1989 edition of *CommuniQue;* "Leadership and Management," published in the summer 1989 edition of the *Quality Management Forum;* "The New Quality Paradigm", published in the September 1989 edition of *CommuniQue.*

Aubrey, C. A. II, and Felkins, P. K. *Teamwork: Involving People in Quality and Productivity Improvement,* ASQC, 1988.

Barra, R. *Putting Quality Circles to Work.* McGraw-Hill, Inc., 1983.

Deming, W. E. *Out of Crisis.* MIT, 1986.

Dumas, R. A. "Organizationwide Quality: How to Avoid Common Pitfalls," published in the May 1989 edition of *Quality Progress.*

Ferrini-Mundy, J., Gaudard, M. Shore, S. D., and Van Osdol, D. "How Quality is Taught Can Be as Important as What is Taught," published in the January 1990 edition of *Quality Progress.*

Godfrey, A. B., "Strategic Quality Management," Part 1 published in the

March 1990 edition of *Quality*, Part 2 published in the April 1990 edition of *Quality*.

Hoernschemeyer, D. "The Four Cornerstones of Excellence," published in the August 1989 edition of *Quality Progress*.

Joiner, B. L. "Continuous Improvement: Putting the Pieces Together," published in the May 1990 edition of the *Quality Management Forum*.

Juran, J. M. *Managerial Breakthrough* McGraw-Hill, Inc., 1964; *Management of Quality* fourth edition. J. M. Juran 1981; "Putting Quality Into the Business Plan," published in Spring 1989 edition of the *Quality Management Forum*.

Juran, J. M., and Gryna, F. M., Jr. *Quality Planning and Analysis*, Second Edition. McGraw-Hill, Inc., 1980.

Kane, E. J. "IBM's Quality Focus on the Business Process," published in the April 1986 edition of *Quality Progress*.

Karabatsos, N "Continuous Improvement—The Challenge for the Nineties," published in the August 1989 edition of *Quality Progress*.

McCabe, W. J."Improving Quality and Cutting Costs in a Service Organization," published in the June 1985 edition of *Quality Progress*.

Melan, E. H. "Process Management in Service and Administrative Operations," published in the June 1985 Edition of *Quality Progress*.

Mohanty, R. P. and Dahanayka, N. "Process Improvement: Evaluation of Methods," published in the September 1989 Edition of *Quality Progress*.

Ott, E. R. *Process Quality Control*. McGraw-Hill, Inc., 1975.

Persico, J., Jr. "Team Up for Quality Improvement," published in the January 1989 edition of *Quality Progress*.

Plsdk, P. E. "Defining Quality at the Market/Development Interface," published in the June 1987 edition of *Quality Progress*.

Roth, W. F., Jr. "Getting Training Out of the Classroom," published in the May 1989 edition of *Quality Progress*.

Sarazen, J. S. "Quality Plan Development: A Key Step Toward Customer Enthusiasm," published in the October 1988 edition of *Quality Progress*.

Scholtes, P. R., and Hacqueford, H. "Beginning the Quality Transformation," published in the July 1988 edition of *Quality Progress*.

Schwinn, D. R., and Schwinn, C. J. "Converting Training into Action," published in the November 1989 edition of *Quality Progress*.

Turnbull, D. M., and Higby, C. W. "Writing Quality Procedures," published in the February 1985 edition of *Quality Progress*.

The American Heritage Dictionary of the English Language, New College Edition. Houghton Mifflin Company, 1981.

ANSI/ASME NQA-1-1986 and 1989 editions, "Quality Assurance Program Requirements for Nuclear Facilities."

ANSI/ASQC Q94-1987 edition, "Quality Management and Quality System Elements—Guidelines."

ANSI/ASQC A3-1987 edition, "Quality Systems Terminology."

ANSI/ASQC Q1-1986 edition, "Generic Guidelines for Auditing of Quality Systems."

Webster's Ninth New Collegiate Dictionary. Merriam-Webster, Inc., 1986.

Index

Achieving Excellence In Business
READER'S UPDATE RECOMMENDATIONS

To: Ken Ebel
  Achieving Excellence
  3101 N. Causeway Blvd., Suite D
  Metairie, LA 70002

Your ideas and examples for improving *Achieving Excellence in Business* will be greatly appreciated and ensure that the book remains a living guide to achieving excellence. This page exists to assist you in providing your information; please do not let it restrict you in any way.

1. Tribulations and trials in implementing the book.

2. Areas needing clarification, and clarification recommendations.

3. Other ideas for improving the book.

Name/position: _____

Business: _____

Address: _____

Phone number: _____

You may use this information in the book: YES   NO

If yes, you may reference my name in the book: YES   NO